AF327125

Resch · Bönisch · Tiyyagura · Furui · Seo · Bez (Eds.)
High Performance Computing on Vector Systems 2006

Michael Resch · Thomas Bönisch · Sunil Tiyyagura
Toshiyuki Furui · Yoshiki Seo · Wolfgang Bez

Editors

High Performance Computing on Vector Systems 2006

Proceedings of the High Performance Computing Center
Stuttgart, March 2006

With 117 Figures, 104 in Color, and 13 Tables

 Springer

Editors

Michael Resch
Thomas Bönisch
Sunil Tiyyagura

Höchstleistungsrechenzentrum
Stuttgart (HLRS)
Universität Stuttgart
Nobelstraße 19
70569 Stuttgart, Germany
resch@hlrs.de
boenisch@hlrs.de
sunil@hlrs.de

Wolfgang Bez

NEC High Performance Computing
Europe GmbH
Prinzenallee 11
40459 Düsseldorf, Germany
wbez@hpce.nec.com

Toshiyuki Furui

NEC Corporation
Nisshin-cho 1-10
183-8501 Tokyo, Japan
t-furui@bq.jp.nec.com

Yoshiki Seo

NEC Corporation
Shimonumabe 1753
211-8666 Kanagawa, Japan

y-seo@ce.jp.nec.com

Front cover figure: The electro-static potential of one pair of $EMIM^+$ (1-ethyl-3-methyl-imidazolium, cation) and $AlCl_4^-$ (anion) projected on an iso-surface of the electron density of the pair in the corresponding ionic liquid $[EMIM^+][AlCl_4^-]$. The potential around the molecules appear as little directional, and this combined with the size of the molecules leads to a low melting temperature.

Library of Congress Control Number: 2006936975

Mathematics Subject Classification (2000): 65-06, 65C20, 68U20

ISBN 978-3-540-47692-4 Springer Berlin Heidelberg New York

Springer is a part of Springer Science+Business Media

springer.com

Typeset by the editors using a Springer TeX macro package
Production: LE-TeX Jelonek, Schmidt & Vöckler GbR, Leipzig
Cover design: WMXDesign GmbH, Heidelberg

Printed on acid-free paper   46/3100/YL - 5 4 3 2 1 0

Preface

With this second issue of "High Performance Computing on Vector Systems - Proceedings of the High Performance Computing Center Stuttgart" we continue our publication of most recent results in high performance computing and innovative architecture. Together with our book series on "High Performance Computing in Science and Engineering'06 – Transactions of the High Performance Computing Center Stuttgart" this book gives an overview of the most recent developments in high performance computing and its use in scientific and engineering applications.

This second issue covers presentations and papers given by scientists in two workshops held at Stuttgart and Tokyo in spring and summer 2006. These workshops are held as part of a collaboration of NEC and HLRS in the "Teraflop Workbench Project" and many of the papers stem from users of the large NEC SX-8 vector systems installed at Stuttgart. At the forefront of research has been the question of how to achieve a high level of sustained performance on vector supercomputers.

The book, however, not only covers application results but you will also find aspects of architectural concepts and innovative systems included. A point of consideration is a comparison of different architectures in terms of performance based on benchmarks. The book hence also covers papers and presentations from speakers that were invited to the workshops coming from outside the traditional vector computing world. With the advent of hybrid systems both in the US and in Japan the importance of such innovative approaches is underlined and future issues of this series will deal also with such hybrid aspects of computer architectures.

The editors would like to thank all authors and Springer for making this publication possible and would like to express their hope that the entire high performance computing community will benefit from it.

Stuttgart, September 2006   *M. Resch*

Contents

Applications II: Molecular Dynamics

Applications III: Environment/Climate Modeling

List of Contributors

M. Auweter-Kurtz
Institut für Raumfahrtsysteme,
Universität Stuttgart,
Pfaffenwaldring 31,
D-70550 Stuttgart
auweter@irs.uni-stuttgart.de

F. Bechstedt
Institut für Festkörpertheorie
und -optik,
Friedrich-Schiller-Universität Jena,
Max-Wien-Platz 1, D-07743 Jena
bechsted@ifto.physik.uni-jena.de

Katharina Benkert
HLRS, Universität Stuttgart,
Nobelstr. 19, D-70569 Stuttgart
benkert@hlrs.de

J. Bernsdorf
CCRLE, NEC Europe Ltd.,
Rathausallee 10,
D-53757 St.Augustin, Germany
j.bernsdorf@ccrl-nece.de

Wolfgang Bez
NEC High Performance Computing
Europe GmbH, Prinzenallee 11,
D-40459 Düsseldorf, Germany
wbez@hpce.nec.com

Arne Biastoch
Leibniz-Institut für
Meereswissenschaften,
Düsternbrooker Weg 20
D-24106 Kiel
abiastoch@ifm-geomar.de

Stefan Borowski
NEC High Performance Computing
Europe GmbH, Heßbrühlstraße 21B,
D-70565 Stuttgart
sborowski@hpce.nec.com

Claus W. Böning
Leibniz-Institut für
Meereswissenschaften,
Düsternbrooker Weg 20,
D-24106 Kiel
cboening@ifm-geomar.de

Thomas Bönisch
HLRS, Universität Stuttgart,
Nobelstr. 19, D-70569 Stuttgart
boenisch@hlrs.de

M. Fertig
Institut für Raumfahrtsysteme,
Universität Stuttgart,
Pfaffenwaldring 31,
D-70550 Stuttgart
fertig@irs.uni-stuttgart.de

F. Fuchs
Institut für Festkörpertheorie
und -optik,
Friedrich-Schiller-Universität Jena,
Max-Wien-Platz 1, D-07743 Jena
fuchs@ifto.physik.uni-jena.de

J. Furthmüller
Institut für Festkörpertheorie
und -optik,
Friedrich-Schiller-Universität Jena,
Max-Wien-Platz 1, D-07743 Jena
furth@ifto.physik.uni-jena.de

Martin Galle
NEC High Performance Computing
Europe GmbH, Heßbrühlstraße 21B,
D-70565 Stuttgart
mgalle@hpce.nec.com

Franz Gähler
ITAP, Universität Stuttgart,
70550 Stuttgart, Germany
gaehler@itap.physik.
uni-stuttgart.de

Stefan Haberhauer
NEC High Performance Computing
Europe GmbH, Heßbrühlstraße 21B,
D-70565 Stuttgart
shaberhauer@hpce.nec.com

S. E. Harrison
Academic Unit of Medical Physics,
University of Sheffield, Glossop
Road, Sheffield, S10 2JF, UK
s.harrison@sheffield.ac.uk

D. R. Hose
Academic Unit of Medical Physics,
University of Sheffield, Glossop
Road, Sheffield, S10 2JF, UK
d.r.hose@sheffield.ac.uk

Barbara Kirchner
Lehrstuhl für Theoretische Chemie,
Universität Bonn,
Wegelerstr. 12, D-53115 Bonn
Kirchner@thch.uni-bonn.de

Pascal Kleijer
NEC Corporation, HPC Marketing
Promotion Division, 1-10,
Nisshin-cho, Fuchu-shi,
Tokyo, 183-8501, Japan
k-pasukaru@ap.jp.nec.com

Hiroaki Kobayashi
Information Synergy Center,
Tohoku University,
Sendai 980-8578, Japan
koba@isc.tohoku.ac.jp

Luis Kornblueh
Max-Planck-Institute for
Meteorology, Bundesstr. 53,
D-20146 Hamburg, Germany
luis.kornblueh@zmaw.de

Uwe Küster
HLRS, Universität Stuttgart,
Nobelstr. 19, D-70569 Stuttgart
kuester@hlrs.de

Peter Lammers
HLRS, Universität Stuttgart,
Nobelstr. 19, D-70569 Stuttgart
lammers@hlrs.de

P. V. Lawford
Academic Unit of Medical Physics,
University of Sheffield, Glossop
Road, Sheffield, S10 2JF, UK
p.lawford@sheffield.ac.uk

R. Leitsmann
Institut für Festkörpertheorie
und -optik,
Friedrich-Schiller-Universität Jena,
Max-Wien-Platz 1, D-07743 Jena
roman@ifto.physik.uni-jena.de

Ralf Messing
IAG, Universität Stuttgart,
Pfaffenwaldring 21,
D-70550 Stuttgart
messing@iag.uni-stuttgart.de

Michael Resch
HLRS, Universität Stuttgart,
Nobelstr. 19, D-70569 Stuttgart
resch@hlrs.de

Ulrich Rist
IAG, Universität Stuttgart,
Pfaffenwaldring 21,
D-70550 Stuttgart
rist@iag.uni-stuttgart.de

Ari P. Seitsonen
CNRS & Université Pierre at
Marie Curie,

4 place Jussieu, case 115,
F-75252 Paris
Ari.P.Seitsonen@iki.fi

S. M. Smith
Academic Unit of Medical Physics,
University of Sheffield, Glossop
Road, Sheffield, S10 2JF, UK

Fredrik Svensson
NEC High Performance Computing
Europe GmbH, Heßbrühlstraße 21B,
D-70565 Stuttgart
fsvensson@hpce.nec.com

Sunil R. Tiyyagura
HLRS, Universität Stuttgart,
Nobelstr. 19, D-70569 Stuttgart
sunil@hlrs.de

Introduction to the Teraflop Workbench Project

The HLRS-NEC Teraflop Workbench – Strategies, Result and Future

Martin Galle[1], Thomas Boenisch[2], Katharina Benkert[2], Stefan Borowski[1], Stefan Haberhauer[1], Peter Lammers[2], Fredrik Svensson[1], Sunil Tiyyagura[2], Michael Resch[2], and Wolfgang Bez[1]

[1] NEC High Performance Computing Europe GmbH
[2] High Performance Computing Center Stuttgart

1 Introduction

This paper is intended to give an overview of the NEC-HLRS cooperation. After a review of the installation phase and a description of the HLRS environment, the major achievements made during the last 12 months within the TERAFLOP Workbench are highlighted. The paper ends with a foresight on future activities.

2 Concept and Targets

Since it's foundation in 2004, the Teraflop Workbench cooperation between HLRS and NEC has successfully provided essential support to the user community in order to enable and facilitate leading edge scientific research. This is achieved by optimizing and adapting existing codes beyond the 1 TFLOP/s threshold and by improving the process work-flow due to the integration of different modules into a "hybrid vector system". The goals of the TERAFLOP Workbench project are:

- Make New Science and Engineering possible with TFLOP/s Sustained Application Performance
- Support the HLRS User Community to Achieve Capability Science with Existing Codes
- Integrate Vector Systems, Linux clusters and SMP Systems towards a "Hybrid Vector System"
- Assess and Demonstrate System Capabilities for Industry Relevant Applications

To reach these goals, NEC and HLRS work together in selected projects with scientific and industrial developers and end users. One member of the

Teraflop Workbench staff is assigned to every project, being in charge of the optimization of the specific application. Furthermore, this member also acts as a contact point for the project partner to the Teraflop Workbench. To optimize the support for the project partner, a frequent exchange of technical issues, experiences and know-how is maintained within the Teraflop Workbench.

The idea behind this unique organization is to combine all experts knowledge, required to set up an efficient environment for leading edge computational science. Application know-how and sound physical background typically is available at the research institutes. HLRS does not only operate the supercomputer environment but also has a long tradition in numerical mathematics and computer science. NEC is able to contribute a deep knowledge of Computer Engineering. The Teraflop staff members have access to internal expertise coming from different specialists groups within HLRS and NEC.

Due to a close collaboration with the hardware and software specialists in Japan, essential input was given for the development of of NEC products, e.g. the SX compiler or the new generations of the SX vector processor. On the other hand, NEC Japan was also able to give valuable contributions to some of the Teraflop Workbench projects.

The Teraflop Workbench is open to new participants. An application has to demonstrate scientific merit as well as suitability and demand for Teraflop performance in order to qualify.

3 The NEC Environment at HLRS

This chapter gives an overview of the NEC installation at HLRS. In Fig. 1 the complete environment is depicted. It consists of 72 nodes SX-8, around 200 PC cluster nodes, each one equipped with two Intel Xeon EM64T (Nocona) 3.2 GHz CPUs.

3.1 Installation and Initiation

The installation of the NEC SX-8 at the HLRS in Stuttgart took place between December 2004 and April 2005. The installation included the hardware and software setup of the TX-7 front end, the IXS and the SX-8 nodes. Additionally, the storage facilities and other peripheral hardware were installed. The installation was carried out by NEC engineers and completed one month ahead of the planning.

Acceptance

Except minor issues, also the acceptance of the system was carried out successfully within the defined time frame. The acceptance tests included:

- Application performance in a single node

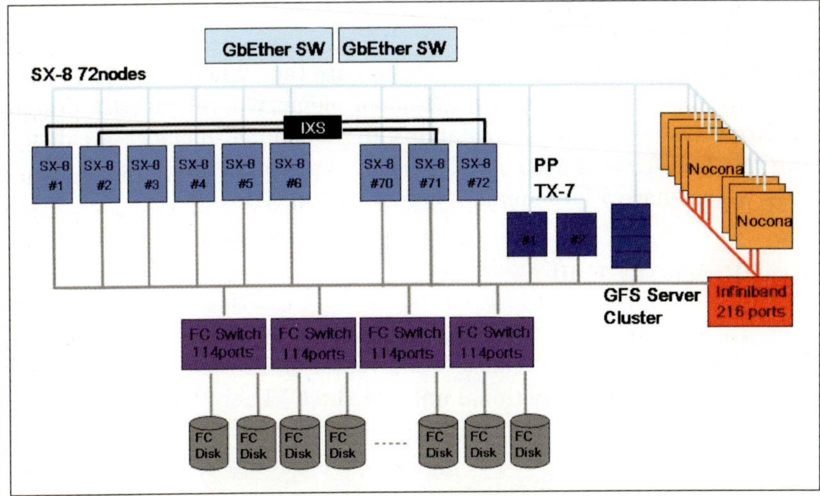

Fig. 1. NEC Installation at HLRS

- Application performance on the complete system
- Memory size
- Memory bandwidth (Single CPU and full node)
- Bisection bandwidth
- MPI bandwidth and latency
- MPI integration of external node (IA64 architecture)
- File System Size
- File System Performance
- Compiler Tests
- Mathematical library availability
- System stability test

Inauguration

The official inauguration ceremonial act was held in presence of the Prime Minister of Baden-Württemberg, Günther Oettinger, the Federal Minister for research and education, Edelgard Bulmahn and NEC Vice President Masahiko Yamamoto.

The scientific user community celebrated the availability of the new environment with a Colloquium "Computational Science on the NEC SX-8". Seven invited talks have been given, demonstrating the capabilities of the NEC SX-8 architecture.

3.2 SX-Compute Nodes

The most important part of the installation are the 72 nodes SX-8. The SX-8 architecture combines the traditional shared memory parallel vector design in Single Node systems with the scalability of distributed memory architecture in Multi Node systems. Each shared memory type single-node system contains 8 CPUs which share a large main memory of 128 GB.

Central Processor Unit

The central processing unit (CPU) is a single chip implementation of the advanced SX architecture. It consists of a vector and a scalar processor. Fig. 2 gives an overview of the functional units of the CPU.

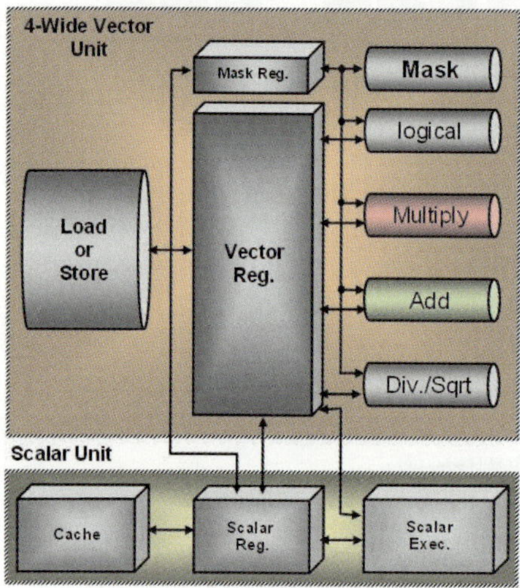

Fig. 2. CPU architecture of SX-8

Vector Unit

A vector unit is equipped with four floating point add/shift and four floating point multiply vector pipelines working in parallel on one single instruction. Additionally, the vector processor also contains four vector logical and four vector divide pipelines. One vector divide pipeline, which also supports vector square root, generates 2 results every second clock cycle. The major clock cycle of the SX-8 is 0.5 nsec, thus the vector floating point peak performance of each processor is 16 GFLOP/s for multiply/add and 4 GFLOP/s for divide/square root.

The vector processor contains 16 KB of vector arithmetic registers which feed the vector pipes as well as 128 KB of vector data registers which serve as a high performance programmable vector buffer that significantly reduces memory traffic in most cases. They are used to store intermediate results and thus avoid memory bottlenecks. The maximum bandwidth between each SX-8 CPU and the shared memory is 64 GB/s. In addition, the CPU is equipped with registers for scalar arithmetic operations and base-address calculations so that scalar arithmetic operations can be performed efficiently.

Scalar Unit

Each CPU contains a 4-way super-scalar unit with 64-kilobyte operand and 64-kilobyte instruction caches. The scalar unit controls the operation of the vector processor and executes scalar instructions. It has 128 x 64 bit general-purpose registers and operates at a 1 GHz clock speed. Advanced features such as branch prediction, data prefetching and out-of-order instruction execution are employed to maximize the throughput. The scalar processor supports one load/store path and one load path between the scalar registers and scalar data cache. Each of the scalar floating point pipelines supports floating point add, floating point multiply and floating point divide operations. The scalar unit executes 2 floating point operations per clock cycle.

Memory Subsystem

The processor to memory port is classified as a single port per processor. Either load or store can occur during any transfer cycle. Each SX processor automatically reorders main memory requests in two important ways. Memory references look-ahead and pre-issue are performed to maximize throughput and minimize memory waits. The issue unit reorders load and store operations to maximize memory path efficiency.

Main Memory Unit

To achieve efficient vector processing a large main memory and high memory throughput that match the processor performance are required. 128 GB DDR2-SDRAM are installed in every node. The bandwidth between each CPU and the main memory is 64 GB/s thus realizing an aggregated memory throughput of 512 GB/s within a single node.

The memory architecture within each single-node frame is a non-blocking crossbar that provides uniform high-speed access to the main memory. This constitutes a symmetric multiprocessor shared memory system (SMP) also known as a parallel vector processor (PVP).

Input-Output Feature (IOF)

Each SX-8 node can have up to 4 I/O features (IOF) which provide an aggregate I/O bandwidth of 12.8 GB/s. The IOF can be equipped with up to 55 channel cards which support industry standard interfaces such as 2 Gb FC, Ultra320-SCSI, 1000base-SX, 10/100/1000base-T. Support for 4 Gb and 10 Gb FC, 10 Gb Ethernet and others are planned. The IOFs operate asynchronously with the processors as independent I/O engines so that central processors are not directly involved in reading and writing to storage media as it is the case in workstation technology based systems.

The SX-8 series offers native FC channels (2 Gb/s) for the connection of the latest, highly reliable, high performance peripheral devices such as RAID disks. FC offers the advantages of connectivity to newer high performance RAID storage systems that are approaching commodity price levels. Furthermore, numerous storage devices can be connected to FC.

SX-8 Internode Communication

Multi node systems of the SX-8 are constructed using the NEC proprietary high speed single-stage crossbar (IXS) linking multiple single node chassis together. The IXS provides very tight coupling between nodes virtually enabling a single system image both from a hardware and a software point of view.

The IXS is a full crossbar providing a high speed single stage non-blocking interconnect. The provided IXS facilities include inter-node addressing and page mapping, remote unit control, inter-node data movement, and remote processor instruction support (e.g. interrupt of a remote CPU). It also contains system global communication registers to enable efficient software synchronization of events occurring across multiple nodes. There are 8 x 64 bit global communication registers available for each node.

Both synchronous and asynchronous transfers are supported. Synchronous transfers are limited to 2 KB, and asynchronous transfers to 32 MB. This is transparent to the user as it is entirely controlled by the NEC MPI library.

The interface technology is based on 3 Gb/s optical interfaces providing approximately $2.7\mu s$ (microsecond) node-to-node hardware latency (with 20 m cable length) and 16 GB/s of node-to-node bi-directional bandwidth per RCU (Remote Control Units). Each SX-8 node is equipped with two RCUs. Utilizing the two RCUs allow for connecting the 72 nodes to a single IXS with a bidirectional bandwidth of 32 GB/s per node.

3.3 EM64T Cluster

For applications which suffer from a poor vectorizability, a PC Cluster is available. It consists of 200 nodes carrying two Intel XEON (Nocona) 3.2 GHz CPUs each.

Compute Nodes

Each processor has a peak performance of 6.4 GFLOP/s and carries 1 MB L2 cache. The XEON architecture supports 64 bits. The processors provide the following performance relevant features:

- super-scalar instruction execution with speculative branching
- out of order execution
- hardware/software prefetching to optimize the instruction execution.
- double speed integer units
- Hyper-Threading
- execution Trace cache
- enhancements in SSE2 and SSE3 execution

The boards which are used in the XEON nodes are equipped with a PCI Express Slot (PCIe x4). In this slot the Infiniband HCAs (Host Channel Adapter) are installed.

Interconnect

The PC Cluster nodes are connected with a Voltaire Infiniband High-speed network. The latency of this interconnect is around 5μs, the bi-directional node-to-node bandwidth is 1800 MB/s.

3.4 File System

On the HLRS system a Global File System (gStorageFS) is installed. It enables the entire Multi Node complex to view a single coherent file system and is working as a client-server concept. The server functionality is implemented on a IA32 based NAS head, managing the I/O requests from the individual clients. The actual I/O however is executed directly between the global disk subsystem and the requesting clients. In future GFS clients will also be installed on the EM64T cluster nodes.

The file system on the NEC SX-8 multi-node system at HLRS is schematically shown in Fig. 3 left.

It consists of 72 S1230 RAID-3 disks. Each RAID has 4 logical units (LUNS) consisting of 8 (+ 1 parity) disks. The NEC SX-8 nodes and the file server are connected to the disks via Fibre Channel switches with a peak transfer rate of 2 Gb/s per port. The file system on the NEC SX-8 cluster, called gStorageFS, is based on the XFS file system. It is a SAN-based (Storage Area Network) file system that takes advantage of a Fibre Channel infrastructure. Large data transfer is performed by using direct client-to-disk I/O. The tested 80 TB file system uses half of the disk resources, namely, 36 S1230 units with 72 controllers. With a total number of 72 FC2 ports at disks and the assumption of 200 MB/s payload on a 2 Gb/s port the I/O FC total limit is calculated at 14.4 GB/s.

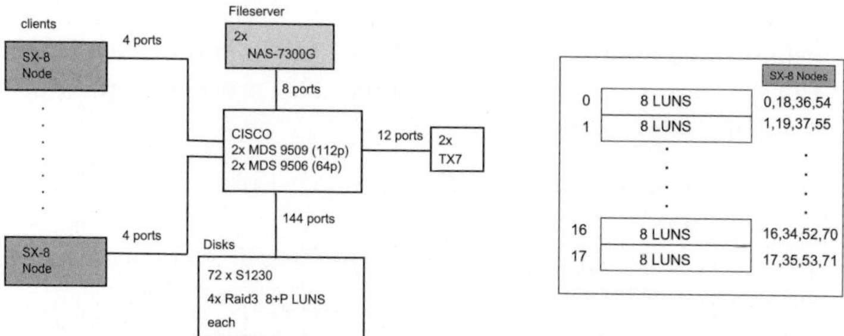

Fig. 3. gStorageFS file system configuration of the NEC SX8 (left). Logical view of file system (right)

The logical view of the file system on the SX-8 cluster is shown in Fig. 3 right. The disks are organized in 18 stripes, each consisting of 8 LUNs. The bandwidth of one LUN is about 100-140 MB/s. A file is created in one stripe, with the location depending on the host creating the file. The bandwidth to access a single file depends on the number of stripes it spans, which is usually one. Fig. 3 right, also shows the assignment of the SX-8 nodes to the stripes. A consequence of this mapping is that if several nodes access the same stripe, they share the bandwidth. Therefore, high aggregate performance can be achieved when multiple nodes access multiple files. Since the striping size is 512 KB, the first block size that makes optimal use of the 8-fold stripe is 4 MB. Larger block sizes increase the efficiency of striping and of access to individual LUNs.

3.5 Operation Data

The HLRS SX-8 system was immediately accepted by the user community. Though the compute power compared to the previous installation consisting of six nodes SX-6 increased by a factor of 12 in the first phase and by a factor of 24 in the second phase, the users were able to scale their problem sizes accordingly without production interrupt. As shown in Fig. 4 the system delivered with 36 available nodes (months April to July) an average performance (floating point operations generated on the complete system divided by calender time, including system downtime and testing), between 1.0 and 1.5 TFLOP/s. After installation of complete system, the average performance increased to more than 2.5 TFLOP/s, which is 20% of the theoretical peak performance. Figure 5 shows the percentage of CPU time related to the available wall clock time. With the availability of half the configuration from April to August, the percentage increased within five months from 40% to almost 80%. Also the complete configuration was quickly adapted by the users. The CPU Time fraction was more than 70% six months after the installation was completed.

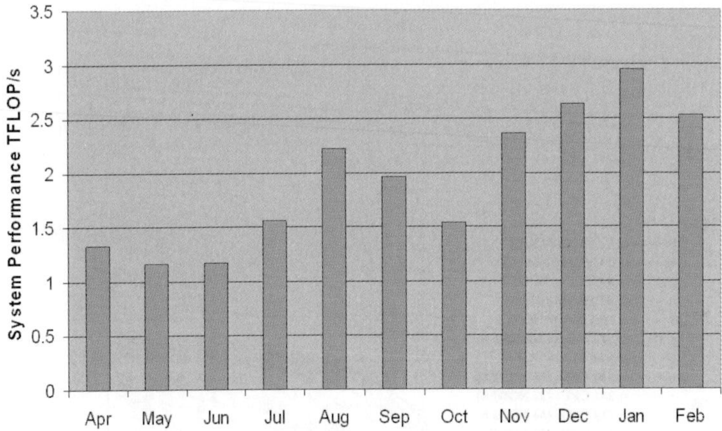

Fig. 4. Operation of the SX-8 at HLRS

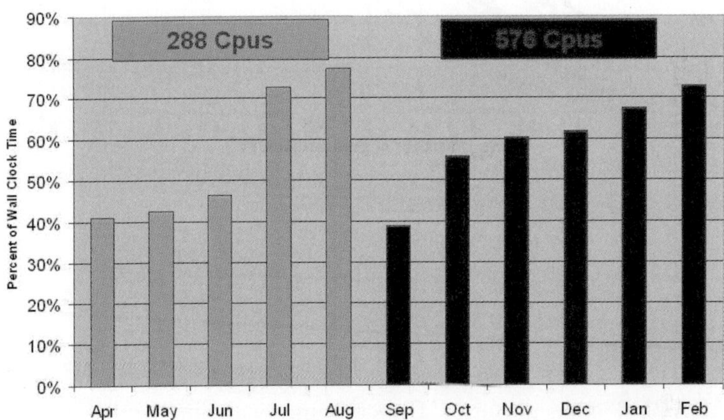

Fig. 5. CPU Time Usage related to available Wall Clock Time

4 Major Achievements

The first three projects (PARAPYR, N3D and FENFLOSS) are already final-
ized while 9 others are ongoing efforts. For two projects there was no activity
during the last six months. Within the near future three new projects will
become active. For the time being, the first target (more than 1 TFLOP/s
sustained performance) is achieved for six codes.

12   Martin Galle et al.

4.1 BEST

The BEST code is a Lattice Boltzmann implementation. It is used for basic
turbulence research. Figure 6 presents the weak scaling behavior of this code
for different problem sizes. For 72 nodes (576 CPUs) the performance reaches
5.68 TFLOP/s for the largest problem size.

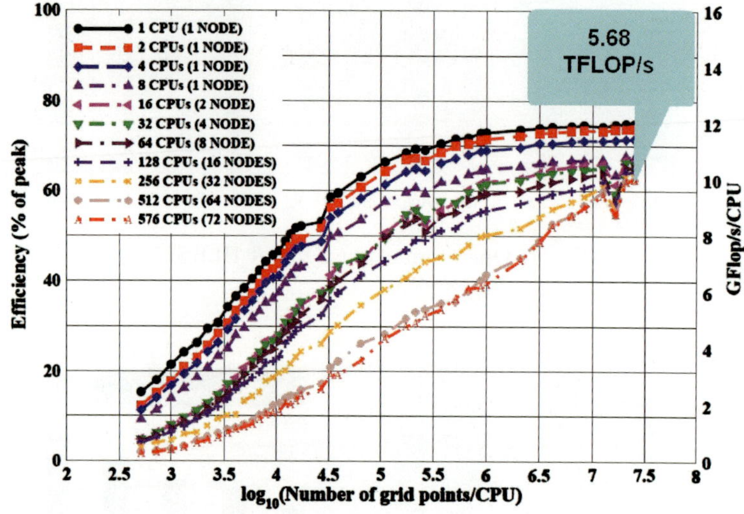

Fig. 6. BEST Performance

4.2 PARAPYR

The PARAPYR code performs Direct Numerical Simulations (DNS). It is
used for simulations of chemical reactions and fluid dynamics during combustion processes. Figure 7 shows the performance estimations and the actual
measured performance. The measured performance is much better than the
estimated performance. This was made possible by additional optimizations
of the code. The performance on 72 node is 4.38 TFLOP/s

4.3 N3D

The N3D code is another DNS application. It is used for basic turbulence
research but also for for drag reduction examinations on aircraft wings. As
depicted in Fig. 8 the measured performance and the estimated performance
show a good overall agreement. On 70 node a performance of 2.68 TFLOP/s
was achieved.

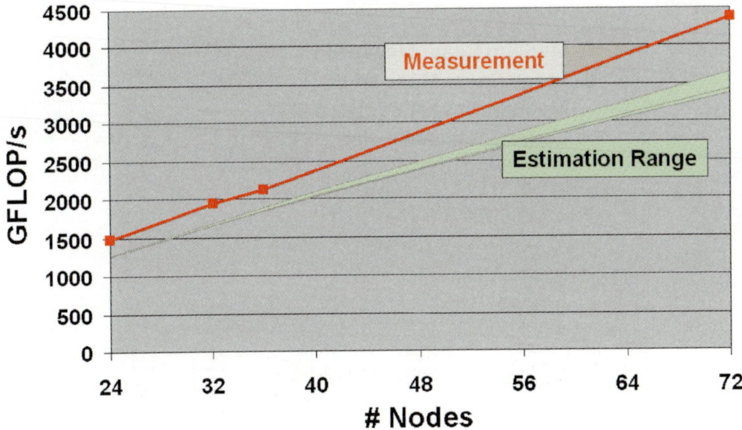

Fig. 7. PARAPYR Performance

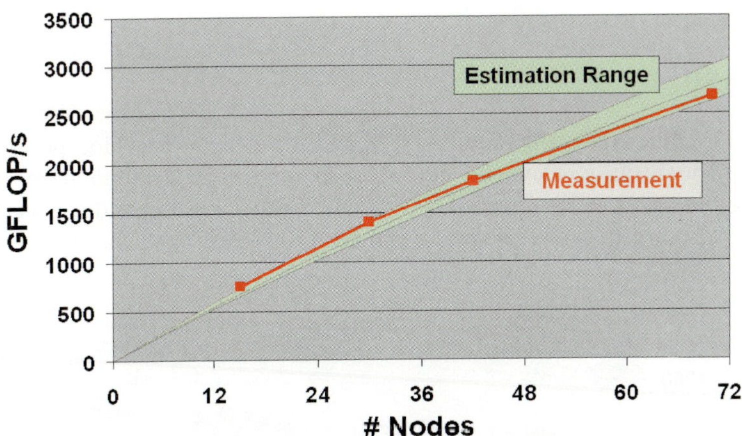

Fig. 8. N3D Performance

4.4 FENFLOSS

FENFLOSS is a Finite Element code employed to simulate flow fields in turbo-machinery related applications. As depicted in Fig. 9 the measured performance for the small case on a large number of nodes does not meet the estimations, while for the large case the measured performance is better than the respective estimation. The scaling of this code is slightly poorer than it was expected, while due to additional optimizations some improvement of the single CPU performance was feasible. The large case reaches 2.59 TFLOP/s on 72 nodes.

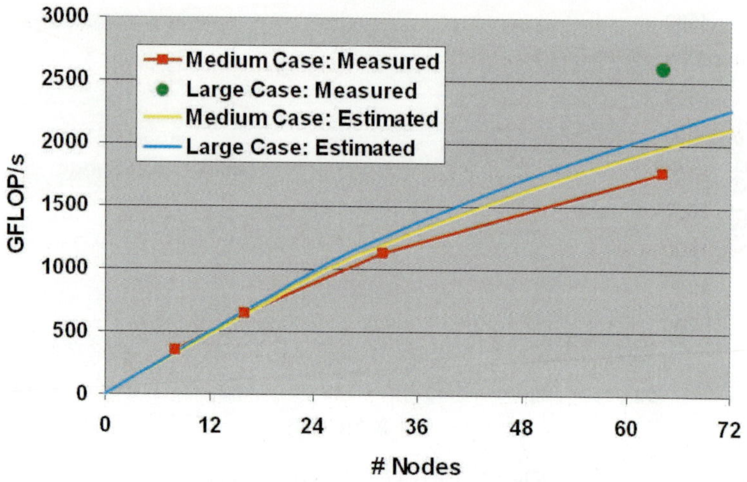

Fig. 9. FENFLOSS Performance

4.5 VASP

VASP (Vienna Ab-Initio Simulation Package) is a package for performing ab-initio quantum-mechanical molecular dynamics (MD) simulations using pseudo-potentials or the projector-augmented wave method and a plane wave basis set. Figure 10 shows the performance of the VASP code for an embedded

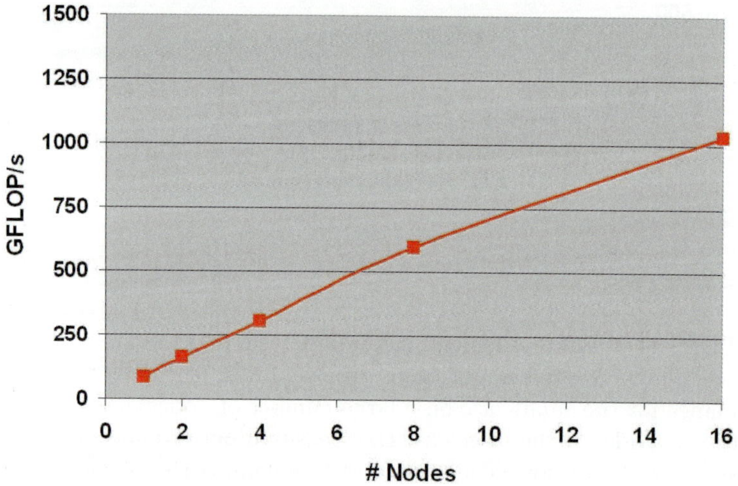

Fig. 10. VASP Performance

quantum dots simulation on the SX-8 system. For this case, the performance on 16 nodes exceeds 1 TFLOP/s.

4.6 CPMD

The Car-Parrinello Molecular Dynamics (CPMD) code is a parallelized plane wave/pseudopotential implementation of Density Functional Theory, particularly designed for ab-initio molecular dynamics. The CPMD code runs very

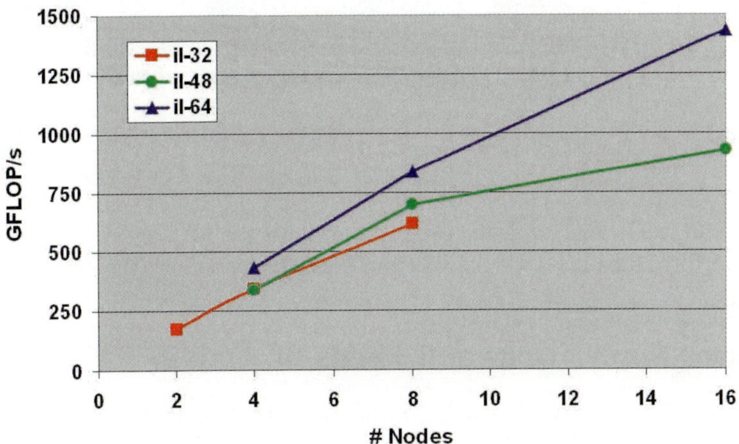

Fig. 11. CPMD Performance

efficient on the SX-8 system. Figure 11 depicts the performance for a simulation of Ionic Liquids (32, 48 and 64 molecules) on up to 16 nodes. The performance on larger configurations increases with the problem size. On 16 nodes, the performance of the medium case is almost 1 TFLOP/s, while for the 64 molecules case 1.4 TFLOP/s are achieved.

4.7 Block Based Linear Iterative Solver BLIS

Within the Teraflop Workbench, the development of a linear iterative solver is carried out. The motivation behind this activity is the limited performance of most of the public domain solvers (like Aztec, PETSc and SPOOLES) on vector machines due to the used data structures, which lead to small vector length or even a large number of small granular calls. Another classical problem lies in the pre-conditioning, where robust algorithms (ILU) perform poorly while simple algorithms (Jacobi) perform relatively well.

The BLIS solver considers the vector architecture by employing a suited Jagged diagonal data structure. Using this data structure, the most time consuming part of the solver, the sparse Matrix-Vector product (MVP), can be

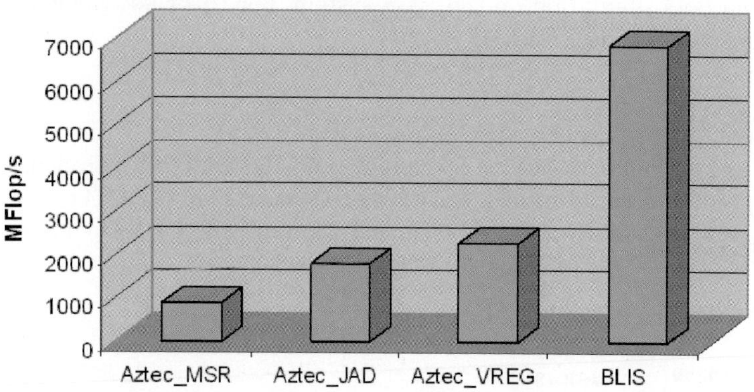

Fig. 12. 1-CPU Performance of Matrix-Vector Product for different solvers

carried out with 45% of peak performance. Figure 12 illustrates the sparse MVP performance per CPU of BLIS in comparison to other solvers.

5 Other Applications and Fields of Activity

Within the Teraflop Workbench, additional codes coming from areas like the classical Molecular or Bio-informatics are analyzed in order to assess the performance potential for these application types. Work-flow optimization is addressed in experiments with heterogeneous MPI applications. Another successful activity of the Teraflop Workbench was the configuration and optimization of the HPCC benchmark on the complete HLRS system.

6 Challenges

During the runtime of the Teraflop Workbench, some problems and shortcomings of the current installation were detected, isolated and described. Most of these could be solved or at least improved by internal efforts or by consulting external expertise within HLRS or NEC. However, some issues still remain:

- Short vector performance and scalability: Small problem sizes are only scalable to a small number of CPUs. In order to use efficiently a large number of nodes, the problem size must be increased significantly. This is not feasible for many types of applications, a better scalability of small problem sizes is desirable.
- Indirect addressing performance: For many classes of applications indirect addressing is an essential programming technique and unavoidable. Though vectorization of indirect addressing data operations is possible, the performance is worse compared to directly addressing code.

- Data Re-Usage: Though the SX-8 has a very high bandwidth to main memory, the performance of many applications are still limited by this bandwidth. A temporary storage of data in cache-like units and re-usage of this data would help to reduce the memory traffic and increase the application performance.

7 Future Activities

7.1 New Application Types

NEC and HLRS intend to broaden the field of application types in the Teraflop Workbench. There are some very interesting areas from which high demands for computational power may arise in the near future. Examples for these application types are:

- Bio-informatics
- Nano-science
- Medical Applications

Another focus will be given to coupled applications, consisting of two or more different programs which run either simultaneously or alternating and which usually have quite different characteristics.

- Fluid and structure
- Simulation and re-meshing
- Ocean and atmosphere
- Aeroacoustics (Noise generation and propagation)
- Flow and chemistry

In general, it cannot be assumed that all modules of a coupled application are best suited to be executed on the same hardware. In order to achieve the optimal performance for those applications, it might be necessary to provide a platform which is based on vector as well as on scalar architecture. This Hybrid Vector Architecture will be discussed in in the following.

7.2 Hybrid Vector Architecture

A suited approach which will become very important in the near future is the Hybrid Vector Architecture. As mentioned in the above section, the idea behind this approach is to provide the best suited architecture for all modules of an application. Typically, but not necessarily, applications which require a hybrid approach are consisting of loosely coupled modules.

As depicted in Fig. 13 the installation at HLRS can already be regarded as a hybrid vector installation, consisting of a strong vector part but also significant computational power being available from scalar based processors (EM64T cluster and the TX-7 with IA64 technology).

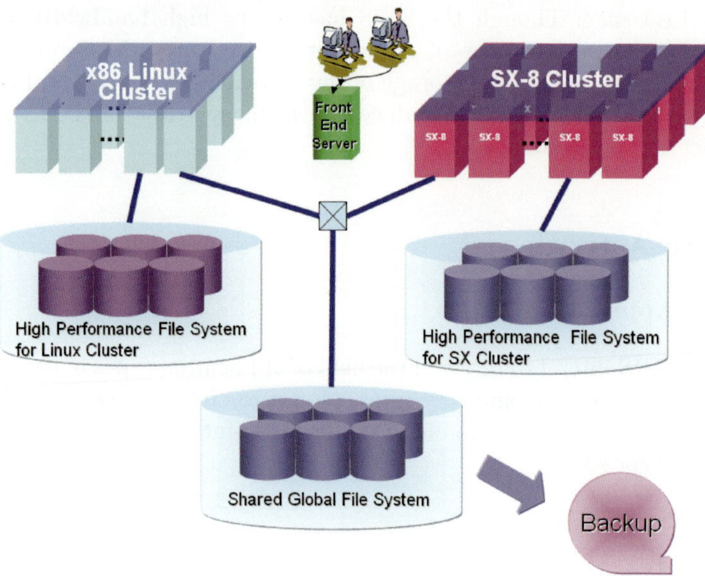

Fig. 13. Hybrid Vector Setup

The coupling of the two architectures is based on a common file system for the time being. This is sufficient for loosely coupled applications, and also an automated work-flow consisting of pre-processing – simulation – post-processing can also be set up straight forward on this type of system.

Future research activities should focus on a closer coupling based on message passing techniques which would allow also closely coupled or even monolithic applications to be executed with a high efficiency on such a system.

8 Conclusions

The installation and acceptance of the SX-8 system at the HLRS in Stuttgart went according and partly ahead to the planning. Even more important is the quick and mainly trouble-free adoption of the system by the user community. One key for this success was the NEC-HLRS cooperation, the "Teraflop Workbench".

The Teraflop Workbench was very successful in the first two years supporting the users in setting up the environment of leading edge scientific work. This includes the optimization and adaptation of user applications on the SX-8 installation, typically resulting in a performance increase by an order of magnitude larger than the previous.

Performance and Libraries

Implication of Memory Performance in Vector-Parallel and Scalar-Parallel HEC Systems

Hiroaki Kobayashi

Information Synergy Center, Tohoku University, Sendai 980-8578, Japan
koba@isc.tohoku.ac.jp

1 Introduction

High end computing (HEC) systems, which are the systems with the highest processing capability, particularly speed of calculation and capacity of memory space, play an important role to accelerate the research and development in the fields of advanced sciences and technologies. The application codes in these fields always need computing power far beyond the peak performance of HEC systems available at the time. The recent report by the HPC task force of Ministry of Education, Culture, Sports, Science and Technology of Japan (MEXT) pointed out that the five important computational science and engineering areas of bio-technology, automobile and aerospace industry, earth and environment, nano-technology, and energy industry request peta-scale computing (10^{15} floating point operations per second) performance for the next generation HEC systems [7]. These never-stop computing demands definitely drive the continuous exponential improvement in performance of the HEC systems that exceeds the pace of Moore's law.

In the last two decades, as one of the seven national supercomputer centers in Japan, we have been running vector-type HEC systems, NEC SX-1, SX-2, SX-3, SX-4, SX-7, and SX-7C, and make them available to academic researchers across the nation. According to the statistics regarding users' SX-7 resource usage in academic year 2005, our vector systems have made a great contribution in a wide variety of leading science and engineering areas, such as perpendicular magnetic recording medium design, high-performance ultra-wide band antennas design, heat-transfer simulation for heat-island analysis of metropolitan area, next generation supersonic transport design, earthquake and sea wave analysis, ozone-hole analysis. The vector system is clearly one of key tools for these areas.

However, when looking at the recent TOP500 ranking [12], the number of vector systems ranked is quite few. In 1993, the systems based on the vector architecture occupied about 67% in ranking, but in 2005 they are rapidly decreased into 3.6% as scalar systems using commodity microprocessors and/or commodity networks become popular in the HPC community. Although the scalar systems show a good performance per price in the certain areas, they are not only solution to satisfy larger demands in high performance computing.

LINPACK, the benchmark used for TOP500 ranking, measures how fast a computer solves dense systems of linear equations $Ax = b$. As LINPACK has high spatial and temporal locality in memory references, and the computation cost is much larger than the communication cost as the the size of data increases, partitioning of the problems becomes easy, and it scales very well in parallel processing. Therefore, more processors (and more data) are provided, more flop/s rates are obtained in the LINPACK test, even though it takes a long time to complete.

In real applications for advanced science and engineering, however, the problems are not so simple. The computations needs a lot of memory references and the communication between processors increases as the the partition of the problems for parallel processing proceeds. Therefore, it would be problematic if TOP500 ranking would become a compass to direct the research and development of HEC systems.

The vector architecture has several advantages compared with the scalar architecture, such as a highly-efficient computing supported by a large number and many kinds of vector pipelines with chaining capability, and high memory bandwidth achieved by a large number of memory banks connected through a high-speed crossbar switch. In addition, our SX-7 provides a large SMP (symmetric multiprocessing) environment with 32 CPUs sharing a single uniform memory space.

The objective of this paper is to clarify the potential of our vector-parallel HEC system, SX-7 and SX-7C using the HPC challenge benchmark suite and several leading application codes in detail, compared with modern scalar HEC systems. So far, some papers have reported the performance comparison between the scalar systems and vector systems [5, 6]. We focus on the effect of memory performance on the sustained system performance when executing the practical applications, and quantitatively clarify how highly-efficient computing of the vector-parallel HEC systems are supported by the well-balanced design regarding memory-bandwidth per flop/s rates, and the number of memory banks.

The rest of the paper is organized as follows. In Sect. 2, after briefly describing the HPC challenge benchmark suite, we present the evaluation results of SX-7 and SX-7C. We show that the SX-7 and SX-7C systems mark the remarkable scores on the memory-related tests of the HPC challenge benchmark compared with modern scalar-parallel HEC systems. In Sect. 3, vector and scalar HEC systems are evaluated using some real applications. In the evaluation, two Itanium2-based scalar systems are selected in addition to the SX-7 and SX-7C. We discuss the the relationship between the cache hit rates and the memory-related processing time in the total execution time of the scalar systems. In addition, we examine the implication of memory-bandwidth per flop/s rates and the number of memory banks for the sustained system performance of the vector systems. In Sect. 4, we report the preliminary performance evaluation of a recently released Intel Montecito core. We believe that the Montecito performance examined in the paper is the first report on a Montecito evaluation using the real simulation codes in the world. Finally, Sect. 5 summarizes the paper.

2 Performance Evaluation of the Vector Systems Using the HPC Challenge Benchmark

2.1 HPC Challenge Benchmark

The HPC challenge benchmark suite [9] has been designed and developed by the DARPA HPCS (High Productivity Computing Systems) program to evaluate high-end computing systems from the wide variety of viewpoints, not only HPL performance in flop/s, which is used for TOP500 ranking, but also memory performance that seriously affects the sustained system performance in real simulation runs. Actually, the HPC challenge benchmark focuses on the memory bandwidth evaluation by using several kernels that have different degree of the spatial and temporal locality of memory references. The suite consists of basically seven tests:

HPL This is the LINPACK TPP benchmark that measures the floating point rate in Tflop/s of execution for solving a linear system of equations on the entire system (named *G-mode*). HPL has a high spatial and temporal locality.

DGEMM This is a subroutine of BLAS (Basic Linear ALgegra Subroutines) and is used to measure the floating point rate of execution of double precision real matrix-matrix multiplication in Gflop/s. DGEMM has a high spatial and temporal locality. DGEMM evaluates the performance of an exclusively running single-MPI process (named *SN-mode*) and per-MPI process performance in embarrassingly parallel execution (named *EP-mode*).

STREAM This is a simple synthetic benchmark program that measures sustainable memory bandwidth in GB/s under performing simple vector operations of copy, scale, sum and triad. STREAM has a high spatial locality, but low temporal locality. STREAM evaluates systems in the SN-mode and EP-mode.

PTRANS This performs parallel matrix transpose through simultaneous communications between pairs of processors, and measures the total communication capacity of the network in GB/s. PTRANS has a high spatial locality, but low temporal locality. PTRANS evaluates systems in the G-mode.

RandomAccess This measures the rate of integer random updates of memory in GUPS (Giga Updates Per Second). As RandomAccess generates highly irregular memory accesses, it has a low spatial and temporal locality. RandomAccess evaluates systems in the G, SN and EP modes.

FFTE This is a kernel program to measure the floating-point rate of execution of double precision complex one-dimensional Discrete Fourier Transform. FFTE has a low spatial locality, but high temporal locality. FFTE evaluates systems in the G, SN and EP modes.

Communication Bandwidth and Latency This measures latency and bandwidth of a number of simultaneous communication patterns (PingPong, natural-ordered rings and random-ordered ring). To measure the bandwidth and latency, 2M bytes and 8 bytes of data are used, respectively.

Through the above seven tests, 28 evaluation metrics in G, SN and EP modes are measured to evaluate high-end systems in terms of computing, memory and network performance.

There are two different runs for benchmarking; baseline run and optimized run. In the baseline run, code modification is not allowed, and the systems are evaluated by using reference implementation codes provided by the HPC challenge committee. On the other hand, in the optimized run, system-specific implementation of codes are allowed to exploit the potential of the tested systems.

2.2 Evaluated Systems and Environments

We evaluate our two modern vector systems, SX-7 and SX-7C that we are running at Tohoku University. A node of SX-7 consists of 32 vector processors sharing a 256GB memory space in an SMP mode [2]. Each vector processor runs at 1.1 GHz and achieves 8.8Gflop/s. Therefore, total SMP performance of an SX-7 node reaches 282.5 Gflop/s. As the memory bandwidth per CPU is 35.3GB/s, the peak memory bandwidth of a node is 1.13TB/s, i.e. 4B/FLOP, which is quite higher than those of modern scalar-parallel cached-based systems.

The vector processor of SX-7 has a vector operation unit and a 4-way superscalar operation unit. The vector operation unit contains four vector pipes (Logical, Add/Shift, Multiply, and Divide) with 144KB vector registers.

The architecture of the SX-7C vector processor is the same as that of SX-7, except for a newly employed SQRT pipe and a 2GHz clock frequency, resulting in a peak performance of 16Gflop/s. A node of SX-7C contains up to 8 processors for SMP with a peak performance of 128Gflop/s and a main memory of 128GB. As SX-7-C also keeps a B/FLOP ratio of 4, the peak memory bandwidth of the node is 512GB/s (64GB/s per processor). Table 1 shows the characteristics of SX-7 and SX-7C.

To evaluate the high SMP capability of SX-7 and SX-7C nodes with a crossbar-connected shared memory, we examine the hybrid parallel processing capability of the systems, in which each MPI process is further processed in an SMP mode, in addition to MPI processing only. Therefore, in the case of the SX-7 evaluation, 32-MPI processing and 2-MPI/16-SMP, which means each MPI process is parallelized using 16 CPUs in an SMP mode, are evaluated. On the other hand, 40-MPI processing and 5-MPI/8-SMP processing on the SX-7C are evaluated.

Table 1. Characteristics of Vector Systems

platform	CPUs /Node	Clock Freq (GHz)	Per CPU		Mem /Node (GB)
			Peak Perf. (Gflop/s)	Memory BW (GB/s)	
SX-7	32	1.1	8.83	35.3(dedicated)	256
SX-7C	8	2.0	16	64.0(dedicated)	128

2.3 Experimental Results and Discussion

In the HPC challenge benchmark, systems are evaluated by using 28 performance metrics of the seven tests in each of base and opt runs, as mentioned in Sect. 2.1. Due to the space limitation, in this paper, we discuss the representative results in each test, and then summarize the overall results in the end of this section. The entire results are available on the HPC challenge benchmark web site at *http://icl.cs.utk.edu/hpcc/hpcc_results.cgi*. In the following discussion, we also use the results of the other systems registered at the HPC challenge benchmark web site for comparison, such as BlueGene/L (BG/L), SGI Altix, IBM p5-575 and p655, Cray XT3 and X1, Sun Fire and Dell PowerEdge.

HPL

As the HPL test evaluates the entire system performance, the absolute performance of SX-7 and SX-7C is quite low compared with the other systems ranked in the HPL test of the HPC Challenge benchmark, for example, 80Tflop/s of the BlueGene/L system with 65K CPUs. However, when looking at the efficiency of the evaluated systems on the HPL test, the SX-7 and SX-7C show the higher efficiency as shown in Fig. 1. The efficiencies of scalar systems are also good because the HPL benchmark has a higher temporal and spatial locality. The HPL is easy to parallelize and scales very well on both vector and scalar systems. A slight difference in efficiency between SX-7 and SX-7C is due to the size of data used. The larger data set given for SX-7C according to the number of CPUs improves its computing efficiency.

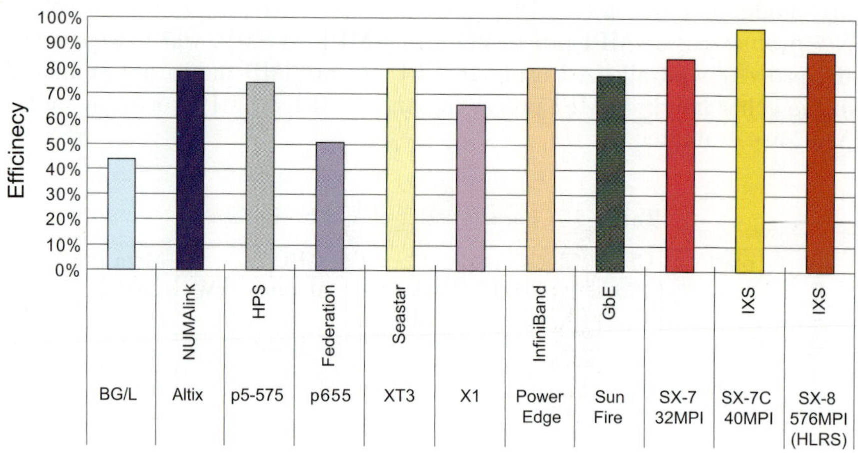

Fig. 1. HPL results in the baseline run

DGEMM

Figure 2 shows the single MPI process performance on the DGEMM, matrix-multiply test. Since DGEMM has also a high spatial and temporal locality, and therefore is cache and vector-load/store friendly, the DGEMM test reflects the peak performance of each system. For example, the performance of SX-8 and SX-7C is two times higher than SX-7, because the formers are running at twice the clock frequency.

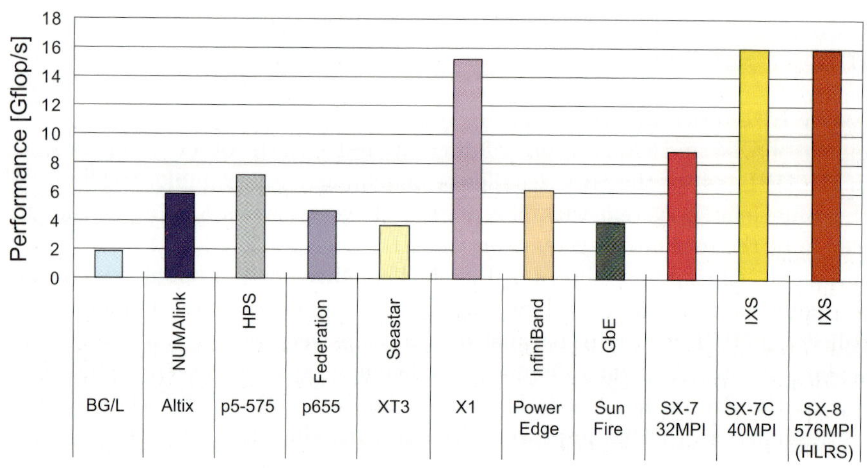

Fig. 2. Single MPI process DGEMM results in the baseline run

In the DGEMM test, parallelization of BLAS by SMP also works very well. Figure 3 shows the effects of hybrid parallel processing on the single-MPI DGEMM test. As the graph shows, the hybrid parallel processing of MPI and SMP enhances the performance as the number of CPUs for SMP increases.

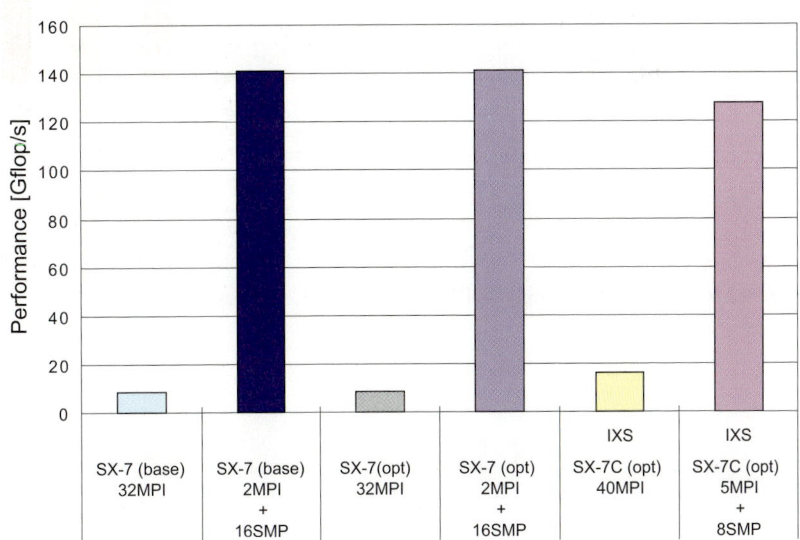

Fig. 3. Effects of hybrid parallel processing on the single MPI DGEMM test

Single MPI Process STREAM

In the case of the memory related tests such as STREAM, the vector systems show the remarkable performance compared with those of the scalar systems. Figure 4 shows the results on the single MPI process STREAM test in the baseline run. The STREAM test measures the effective memory bandwidth of the systems in GB/s. As the figure shows, the vector systems such as our SX-7 and SX-7C, SX-8 of HLRS (High Performance Computing Center Stuttgart of University of Stuttgart, Germany), and CRAY X1 outperform the scalar systems by a factor of 10 or more. The higher memory bandwidth of the SX-7 system is made available by high-memory bandwidth of 35.3GB/s (1.13TB/s per node), and a large number of memory banks of 512 banks per CPU (16K banks in a single node). It can achieve the 4B/flop ratio. The high performance of the vector systems on the stream test completely reflects their higher memory bandwidth.

Figure 5 shows the performance in the optimized runs. BlueGene/L uses some custom code to enhance the performance to 2GB/s, but it is still 25 to 48 times lower than those of the SX systems.

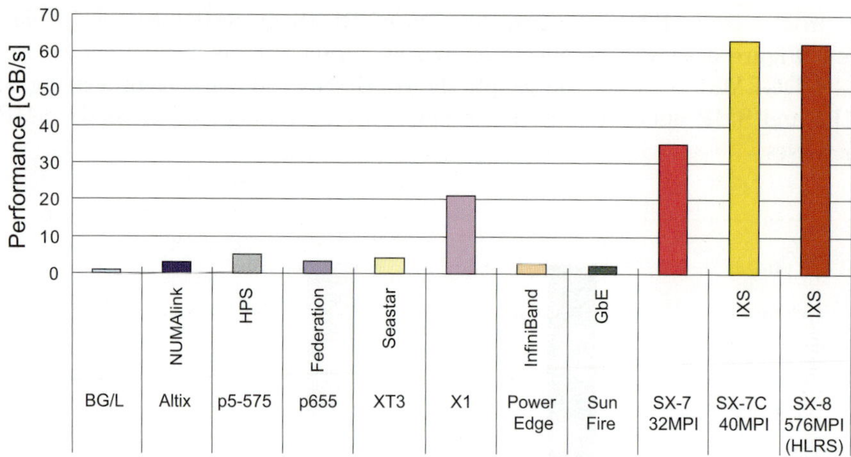

Fig. 4. Results on the single MPI process STREAM test in the baseline run

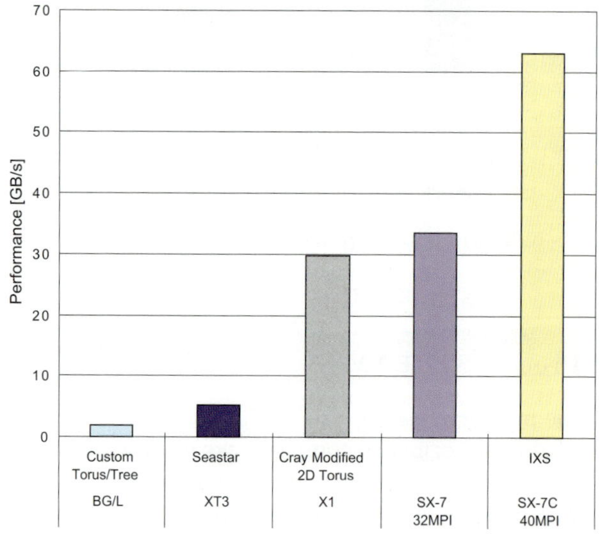

Fig. 5. Results on the single MPI process STREAM test in the optimized run

PTRANS

The PTRANS test is a parallel matrix transpose program, and is used for
the evaluation of inter-CPU communication performance in GB/s. This test,
requiring repeated reads, stores, and communications among processors, is
a measure of the total communication capacity of the interconnect. The
PTRANS test evaluates the entire system performance in the G-mode. Figure 6 shows the results on the PTRANS test in the baseline run. In the
PTRANS test, XT3 shows the outstanding performance of 1.8TB/s using

more than 10K CPUs, which is the first system that exceed the TB/s performance. The SX systems with the small number of CPUs (32 CPUs of SX-7 and 40 CPUs of SX-7C) are not competitive on the entire system performance test, compared with the other systems with a larger number of CPUs. However, the vector systems have a highly efficient computing capability. For example, the HLRS SX-8 system using 576 CPUs outperforms the scalar systems with thousands CPUs. The highly efficient computing of the vector systems are also confirmed through the normalized PTRANS in bytes/Kflop, which are obtained by dividing the PTRANS performance by the system peak performance as shown in Fig. 7. The SX systems show quite higher performance than the other systems.

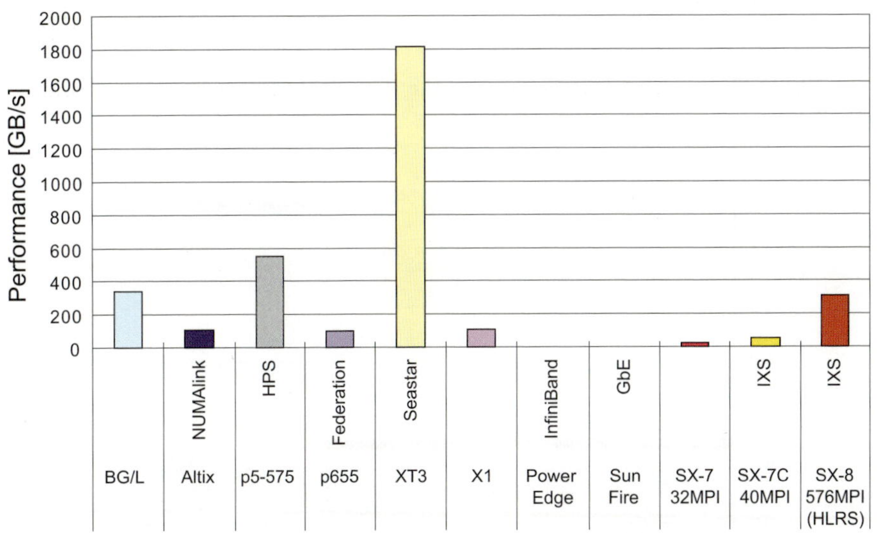

Fig. 6. Results on the PTRANS test in the baseline run

For the optimized run of PTRANS, we carefully tuned the parameters for the buffer size of MPI data transfer. As shown in Fig. 8, the buffer size for MPI data transfer is fixed for the certain data size. Therefore, by selecting the data size best suited for the given buffer size, the PTRANS performance is maximized accordingly as shown in Fig. 9.

Single MPI Process Random Access

The random access benchmark test checks performance in moving individual data rather than large arrays of data. Moving individual data quickly and well means that the computer can handle chaotic situations efficiently. Figure 10 shows the results on the random access test in the baseline run, measured

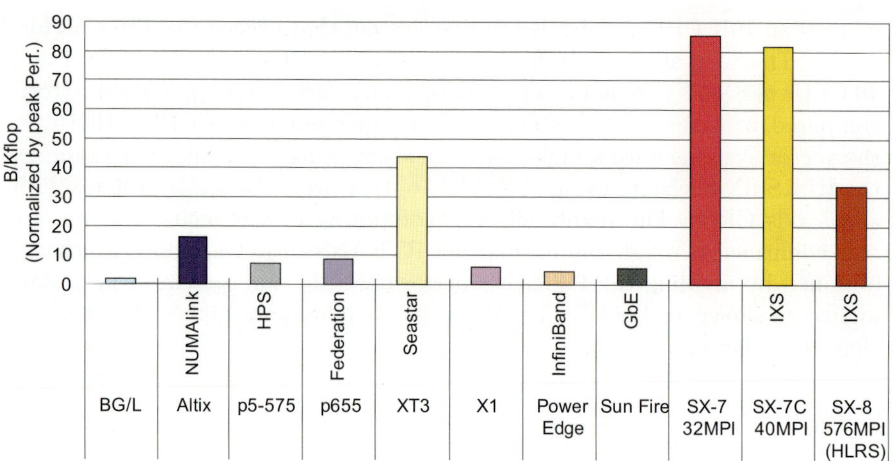

Fig. 7. Normalized performance on the PTRANS test in the baseline run

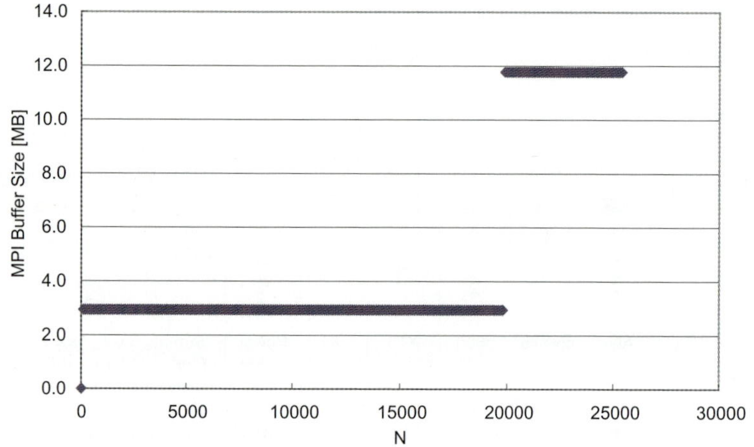

Fig. 8. MPI buffer size vs. data size(N)

in the unit of giga updates per second. As this test enforces highly-irregular memory accesses on the systems, the cache memory of the scalar systems does not work efficiently, and the scalar systems show the very low performance on this test. On the other hand, the higher memory bandwidth of the vector systems contributes to the achievement of the remarkable performance on the random access test. The performance of the SX systems is 40 to 60 times higher than those of the scalar systems. Twice higher clock frequency of the SX-7C leads to a 50% increase in random access performance compared with SX-7. A 15% down of SX-8 from SX-7C may be due to the compiler performance, because they have the same peak performance, but use different versions of compilers.

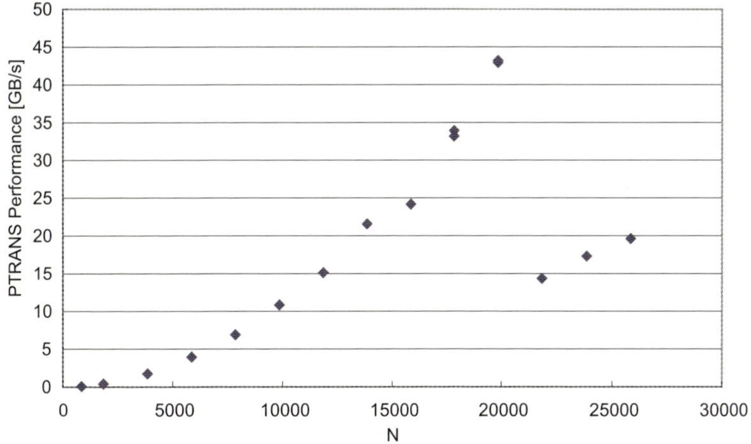

Fig. 9. Relationship between PTRANS performance and data size

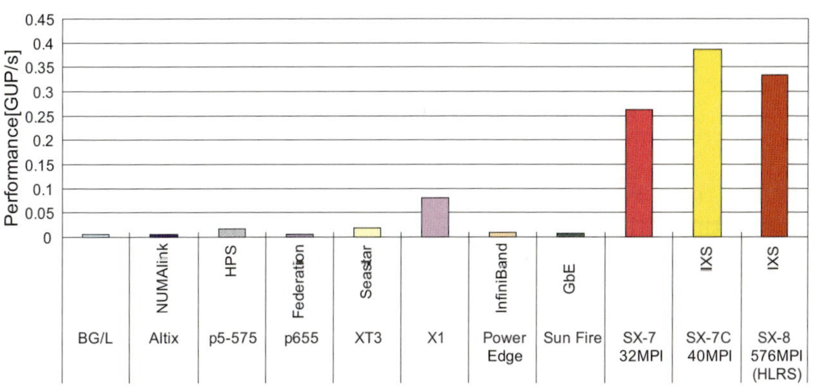

Fig. 10. Single MPI process random access results in the baseline run

For the optimized run, we tune the code to increase the vector length from 128 to 256, and this leads to a 50% improvement as shown in Fig. 11. Cray also enhanced random access performance in the optimized run by changing vector length and adding concurrent directive, but still remains the half of SX-7C performance.

Single FFTE

The FFT code of the HPC challenge benchmark is basically designed for cached-based system, and parameter $L2SIZE$ is provided to specify the cache size of individual systems. However, in the baseline run, as code modification is not allowed, the performance of the FFT run on the evaluated systems is not so good as shown in Fig. 12. Regarding the vector systems, vectorization

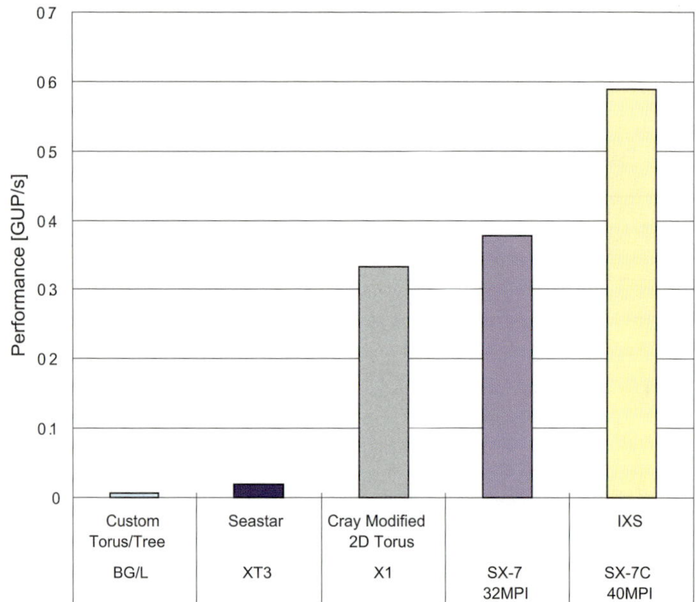

Fig. 11. Performance comparison in the optimized run

of the FFT code is essential to extract their potentials. Figure 13 shows the results in the optimized run, where system-specific optimization is applied in individual executions of the vector and scalar systems. We point out that the vectorization for SX-7 and SX-7C is quite effective and remarkably improved the performance by a factor of 10. In addition, the hierarchical parallel processing of MPI and SMP works very well on the single FFTE test. The speedup ratios of hybrid to non-hybrid are 3.4 in SX-7 (baseline run), 6.1 in SX-7 (optimized run), and 4.8 in SX-7C (optimized run) as shown in Fig. 14.

Latency and Bandwidth

Figures 15 and 16 show the latency and bandwidth of the network systems. Among them, Altix shows the excellent network performance on both latency and bandwidth tests. Generally, custom networks show higher performance compared with commodity networks. The exception is BlueGene/L. Its latency is small, but the bandwidth is as low as that of the giga-bit ether network. This means that BlueGene/L can efficiently handle a large number of short messages between CPUs. InfiniBand shows the similar performance of BlueGene/L. As giga-bit ether networks have long latency and low bandwidth, they are not adequate for the execution of parallel programs with a large number of data communications.

The SX vector systems show quite higher bandwidth compared with the scalar systems. Of course, the internal crossbar switch of SX-7 achieves high

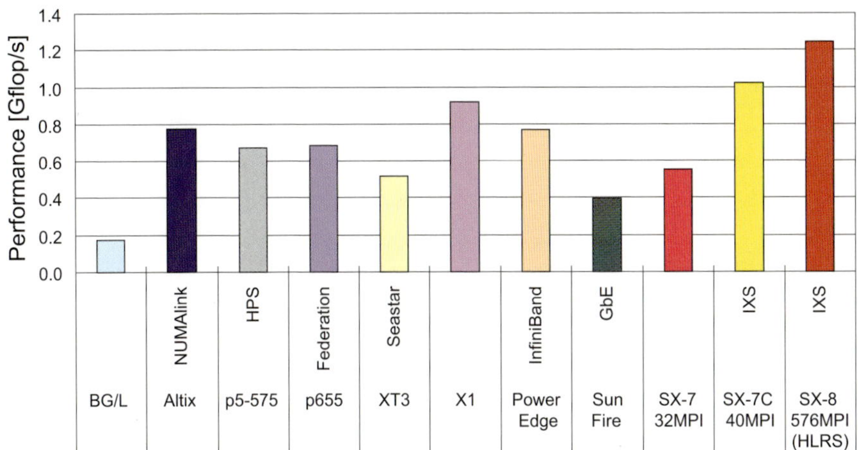

Fig. 12. Single FFTE results in the baseline run

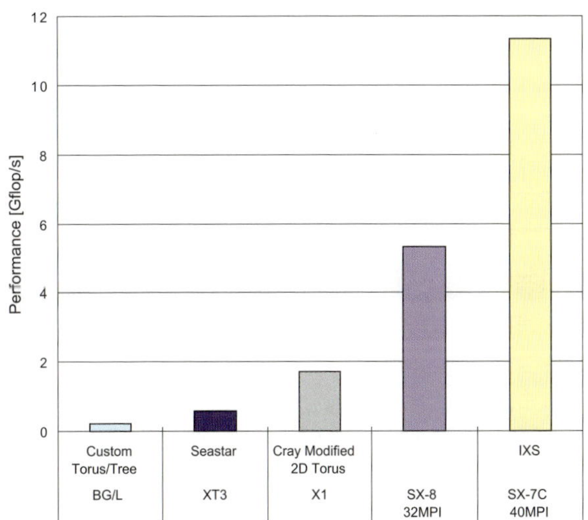

Fig. 13. Single FFTE results in the optimized run

network performance, but SX-7C and SX-8 with inter-node IXS networks also outperforms the other custom networks, for example, Cray X1. From these observation, the vector systems can achieve very fast transmission of large data set between CPUs by taking advantage of their high memory bandwidths.

Overall Results

Figures 17 and 18 show the overall results of some representative systems on all the 28 tests in the baseline run and the optimized run, respectively. The

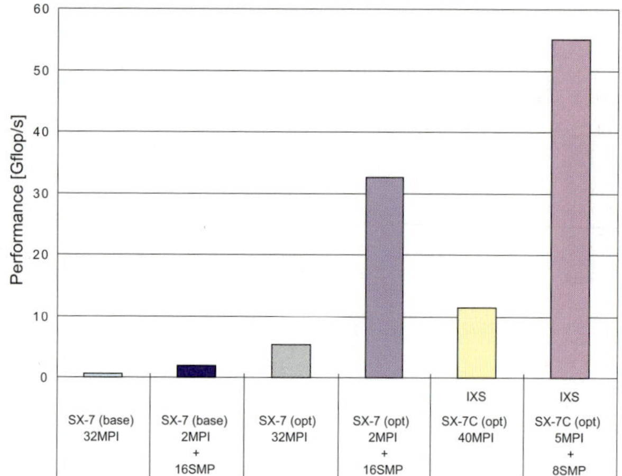

Fig. 14. The effect of hybrid parallel processing in the single FFT test

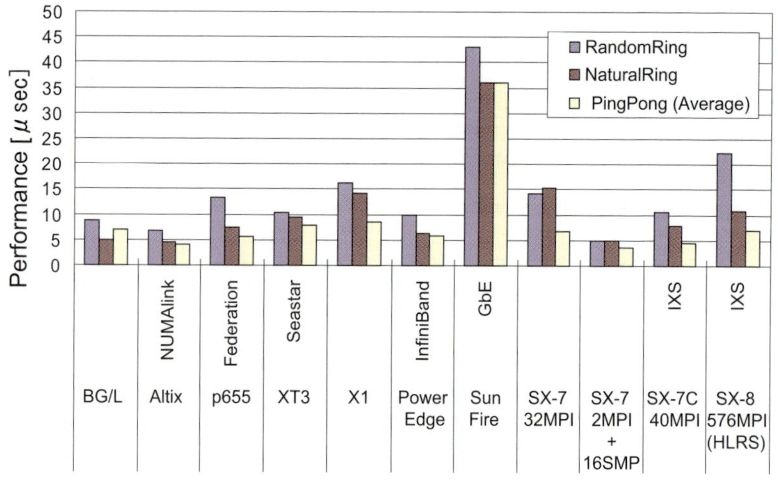

Fig. 15. Latency in the baseline run

outermost circle on 100% of each figure is the position for the system that marks the best performance, i.e., *Top 1 score*, on each test. The performance of the other systems are normalized by the best performance, and ranked accordingly. The detailed results are available on the HPC Challenge benchmark web site at *http://icl.cs.utk.edu/hpcc/hpcc_results_all.cgi*.

As the figures show, the vector systems can achieve quite impressive performance on the memory-related tests, although the number of CPUs of the vector systems is relatively small. On the other hand, the scalar systems with a large number of CPUs show the best performance on the global tests that

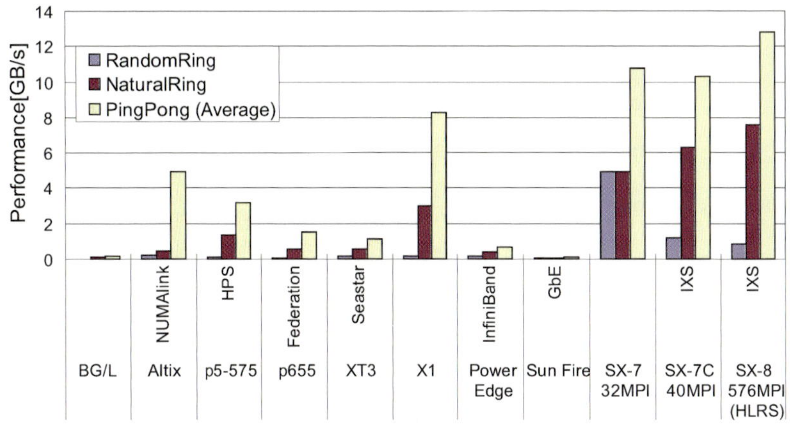

Fig. 16. Bandwidth in the baseline run

can be increased as the number of CPUs increases. For example, in the baseline run, the hybrid parallel processing of MPI and SMP on SX-7 marks 11 top-scores on the 28 test, i.e., 8 on single and embarrassingly parallel (EP) memory-related tests, and 3 on the single and EP flop/s-related tests. SX-8 of HLRS marks top-1 score on max PingPong bandwidth. The BlueGene/L with 65K CPUs is ranked at top-1 on the global HPL and global FFT tests.

As the radar charts of the HPC challenge benchmark results suggest, individual HEC systems have their own characteristics. In the applications in which their temporal/spatial locality is not so high, the sustained system performance in execution of the application codes is seriously affected by memory performance, and therefore, we think that an excellent balance of memory bandwidth per flop/s, 4 bytes/flop of the SX systems, is an important factor to realize highly efficient supercomputing.

3 Performance Evaluation of the Vector and Scalar Systems Using Real Application Codes

3.1 Evaluation Environment

The true value of high-end computing systems is neither their peak performance listed on the spec sheets nor best performance on the benchmark tests using synthesized kernels, but how much sustained system performance is obtained in execution of real simulation codes for advanced science and engineering. To increase the sustained performance, we think that memory performance is a key factor because most real simulation codes need highly irregular memory accesses [10]. US HECRTF (High-End Computing Revitalization Task Force) pointed out in 2004 the divergence problem that means there is an

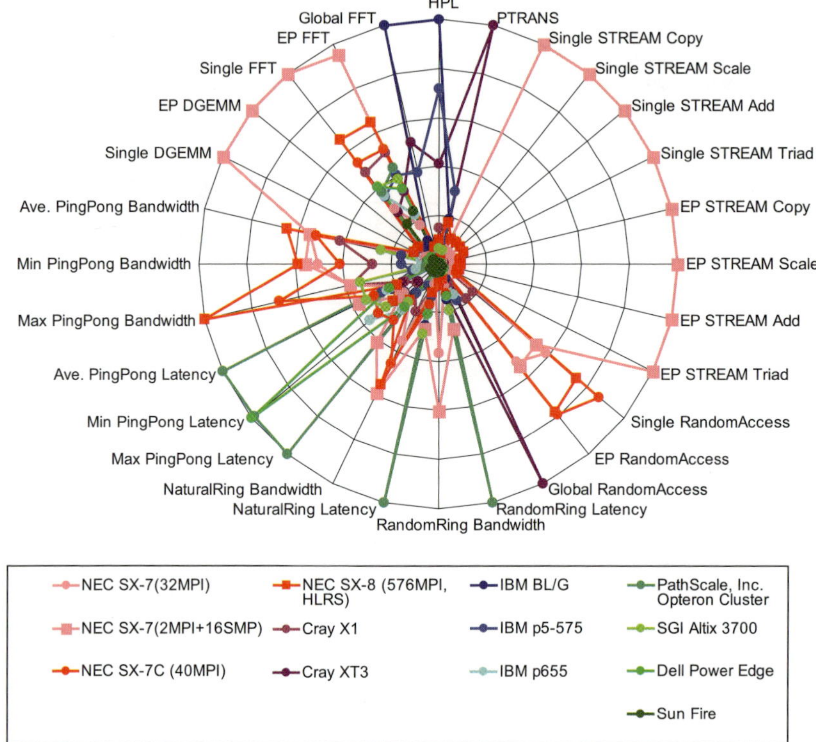

Fig. 17. Overall Results on the HPC challenge benchmark tests in the baseline run

increasing gap between the peak performance and the sustained performance
for major high-end computing centers in the US. Thanks to the Moore's law,
the processor performance goes up rapidly at the historical rate of 50% per
year, but the memory bandwidth improves at 5% only. This imbalance produces the disappointing rise in sustained system performance.

To clarify the effect of memory architectures and their performance of high-
end computing systems on the sustained system performance, we evaluate
HEC systems using real simulation codes that are actually used in the leading
science and engineering areas. We pick up two modern scalar systems, NEC
TX-7/i9510 and SGI Altix 3700, in addition to SX-7 and SX-7C for this
purpose.

The TX7/i9510 and Altix systems are equipped with Itanium 2 processors
running at 1.6GHz, and have the ccNUMA (Cache Coherent Non-Uniform
Memory Access) architecture. A node of the TX7/i9510 has eight cells and
crossbar network modules [13]. Each cell contains four Intel Itanium2 processors and a 32GB main memory, which are interconnected by a 6.4GB/s bus.
A cell card of Altix3700 consists of four Intel Itanium2 processors, main memory, and two controller ASICs called SHUB. A SHUB connects two processors

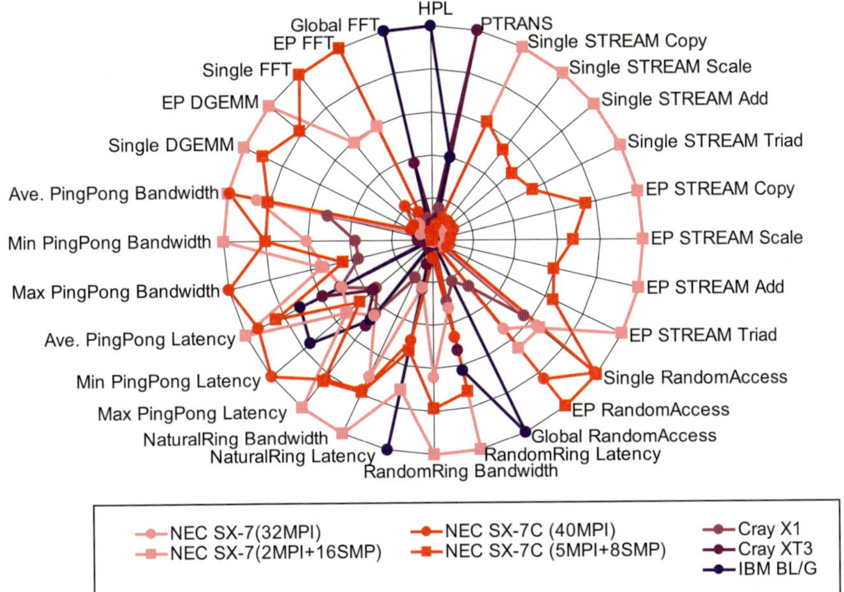

Fig. 18. Overall Results on the HPC challenge benchmark tests in the optimized run

and memory at a 6.4GB/s. The Altix interconnect is called the NUMA link, custom network in a fat-tree topology. The Itanium2 has a three-level on-chip cache memory. Table 2 summarizes the characteristics of the scalar systems disused in this paper.

The big difference between the vector systems shown in Table 1 and the scalar systems is memory performance. The vector systems have 5 to 10 times faster memory bandwidth per CPU. In the following discussion, we examine how memory performance affects the sustained system performance in execution of real simulation codes.

Table 2. Characteristics of the scalar systems

System	CPUs /Node	Clock Freq (GHz)	Per CPU			Mem /Node
			Peak Perf. (Gflop/s)	Memory BW (GB/s)	L3 Cache (MB)	(GB)
TX7/i9510	32	1.6	6.4	6.4(shared)	9	128
Altix3700	64	1.6	6.4	6.4(shared)	6	128

Five simulation codes of leading applications from three areas in scientific computing are used to compare the sustained performance of SX-7 and SX-7C with that of TX7/i9510 and Altix3700. Table 3 shows the summary of five applications. The applications have been developed by researchers of Tohoku University, and actually used in each research area.

EM Scatter

This code is for a simulation of an array antenna, named *SAR-GPR (Synthetic Aperture Radar-Ground Penetrating Radar)*, which detects anti personnel mines in shallow subsurface [11]. This simulation method is the three dimensional FDTD (Finite Difference Time Domain) method with Berenger's PML (Perfectly Matched Layer) [3]. The FDTD is a way to describe a finite difference form of Maxwell's equations, and obtains electromagnetic field in a simulation space. The simulation space consists of two regions; air-space and subsurface space with PML of 10 layers. The performance of this code is primarily determined by the performance of electromagnetic field calculation processes. The basic computation structure of the processes consists of triple-nested loops with non-unit stride memory accesses. The ratio of the calculation cost to the total execution is 80%. The innermost loop length is 500+.

Antenna

The antenna code is for studying radiation patterns of an Anti-Podal Fermi Antenna (APFA) to design high gain antennas [14]. The simulation consists of two sections, a calculation of the electromagnetic field around an APFA using the FDTD method with Berenger's PML, and an analysis of the radiation patterns using the Fourier transform. The performance of the simulation is primarily determined by calculations of the radiation patterns. The ratio of its calculation cost to the total execution is 99%. The computation structure of the calculations is triple-nested loops; the innermost loop is a unit-stride loop, and its length is 255. On Itanium2, the innermost loop is executed on the caches. The ratio of floating-point operations to memory references in the innermost loop is 2.25. Therefore, this code is computational-intensive and cache-friendly.

Combustion

This code realizes direct numerical simulations of two-dimensional Premixed Reactive Flow (PRF) in combustion for design of plane engines [4]. The simulation uses the 6th-order compact finite difference scheme and the 3rd-order Runge-Kutta method for time advancement. The performance of this code is primarily determined by calculations of derivations of physical equations. The ratio of its calculation cost to the total execution is 50%, and the rest of the cost has been distributed among various routines. The code has doubly nested loop; the loop of x-derivations induces unit-stride memory accesses, and the loop of y-derivations induces non-unit-stride memory accesses. The length of each innermost loop is 513.

Heat Transfer

This simulation code realizes direct numerical simulations of three-dimensional laminar separated flow and heat transfer on plane surfaces [8]. The finite-difference forms are the 5th-order upwind difference scheme for space derivatives and the Crank-Nicholson method for a time derivative. The resulting finite-difference equations are solved using the Simplified Marker And Cell (SMAC) method. The performance of the code is primarily determined by calculations of the predictor-corrector methods. The ratio of its calculation cost to the total execution is 67%, and the rest of the cost has been distributed among various routines. The code has triple-nested loops; the innermost loop needs unit-stride memory accesses, and its length is 349.

Earthquake

This code uses the three-dimensional numerical plate boundary models to explain an observed variation in propagation speed of post-seismic slip [1]. This is a quasi-static simulation in an elastic half-space including a rate and state-dependent friction. The performance of the simulation is primarily determined by a process of thrust stress with the Green function. The ratio of its calculation cost to the total execution is 99%. The computation structure of the process is a doubly nested loop that calculates a product of matrices. The innermost loop needs unit-stride memory accesses and its length is 32400.

Table 3. Characteristics of Simulation Codes

Area	Simulation Code Name	Method	Subdivisions
Electro-Magnetic Analysis	*EM Scattering* Analysis of antenna radiation in a bore hole	FDTD	50x750x750
Electro-Magnetic Analysis	*Antenna* Electromagnetic wave analysis	FDTD	612x105x505
CFD/ Heat Analysis	*Combustion* Instability analysis of 2-dimensional premixed reactive flow in combustion	DNS	513x513
CFD/ Heat Analysis	*Heat Transfer* Analysis of separated flow and heat transfer	SMAC	711x91x221
Earth-quake Analysis	*Earthquake* Earthquake analysis of seismic slow slip on the plate boundary	Friction Law	32400x32400

The experiments in this work measure the performance of the original source codes, which were developed for SX-7, with optimizations of the compilers. The compiler's high-level optimizations (SX: -C hopt, TX and Altix: -O3) including inline expansion of subroutines are applied. We used NEC compiler for the Intel Itanium2 on the TX7 and Altix to evaluate the performance under the same level optimizations. On SX-7 and SX-7C, the five applications are vectorized by compiler. The applications are automatically parallelized by compiler on four studied systems. For these experiments, the period for initialization in execution is skipped to measure the performance in the steady state. All the performance statistics of four studied systems were obtained using the NEC compiler option *ftrace*.

3.2 Experimental Results and Discussion

Figure 19 shows the efficiency of sustained performance obtained by running the five simulation codes on the individual systems. As the figure shows, the vector systems keep a higher efficiency of 40%, but that of the scalar systems is 10% or lower across the five simulation codes. To clarify the reason for these quite different results, we examine the memory processing time in the total execution time of the codes on the four systems. Figure 20 presents the ratio of memory time to the total processing time. Here, memory time is the time for memory reference that cannot be hidden by computations, and is exposed in the total execution time. In the vector systems, the ratio of memory processing to the total processing time is quite low, around 20% to 30% in the EM-scatter, Combustion and Heat-Transfer, and about 2% in Antenna and Earthquake. However, on the scalar systems, the processing time for memory accesses becomes dominant in the total execution time of all the five simulation codes, for example, 95% in EM-Scatter. Even in the base case for the scalar systems, where the on-chip cache hit achieves 99%, 20+% of the total processing time is spent for the memory reference. These large memory overheads result in the poor efficiency in execution of the practical application codes.

Figure 21 shows the scalability of the four systems when using 32 CPUs. Because the maximum number of CPUs per node of SX-7C is 8, we discuss the performance of the three systems (SX-7, TX7/i9510, and Altix3700) In Combustion, the scalability is limited because of its low parallelism. In the case where the memory processing time is relatively low, e.g., Antenna, the performance of the scalar and vector systems scales well. SX-7 still outperforms the other systems across all of the applications. Regarding the scalar systems, Altix shows the better scalability compared with TX7/i9510. This is because there are a lot of access contentions on the TX7 memory bus on which twice the CPUs are connected compared with Altix. Let us take a close look at the results of EM-Scatter.

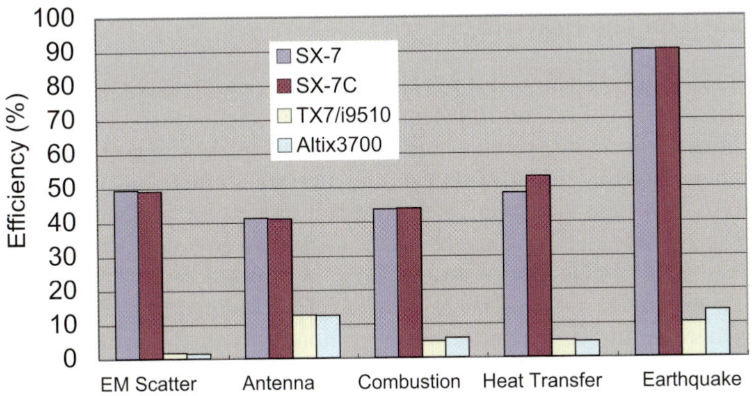

Fig. 19. Efficiency of the four system for the five simulation codes

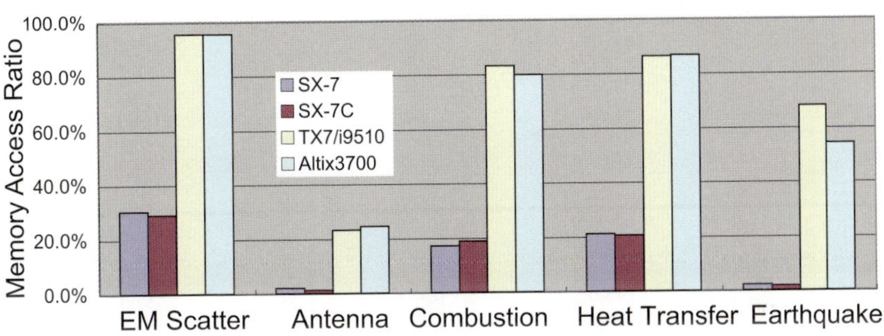

Fig. 20. Memory access ratios to the total execution time

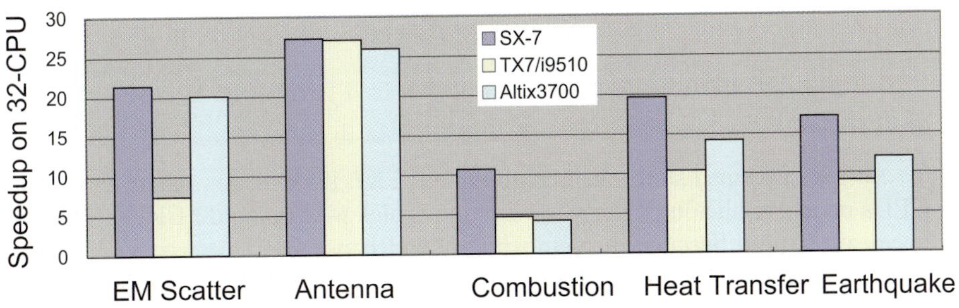

Fig. 21. Scalability

Implication of Memory Performance for Sustained System Performance on the Scalar Systems

Figures 22, 23, and 24 show the relative performance in execution of EM-Scatter normalized by TX7/i9510 performance, the efficiency to the peak performance of the individual systems, and the scalability. As the vector systems achieve higher efficiency in performance, they overwhelmingly outperform the scalar system from the viewpoint of the sustained system performance. For example, the relative performance of SX-7 reaches 43 by one CPU, and 126 by 32 CPUs. SX-7C also shows an excellent performance, but the relative performance gradually declines as the number of CPUs increases. This means more slow increase in SX-7C performance compared with the growth rate in TX7/i9510 performance. This is due to the lack of memory banks of SX-7C as the number of CPUs goes up, although the system keeps 4B per flop rate irrespective of the number of CPU used. This will be discussed later.

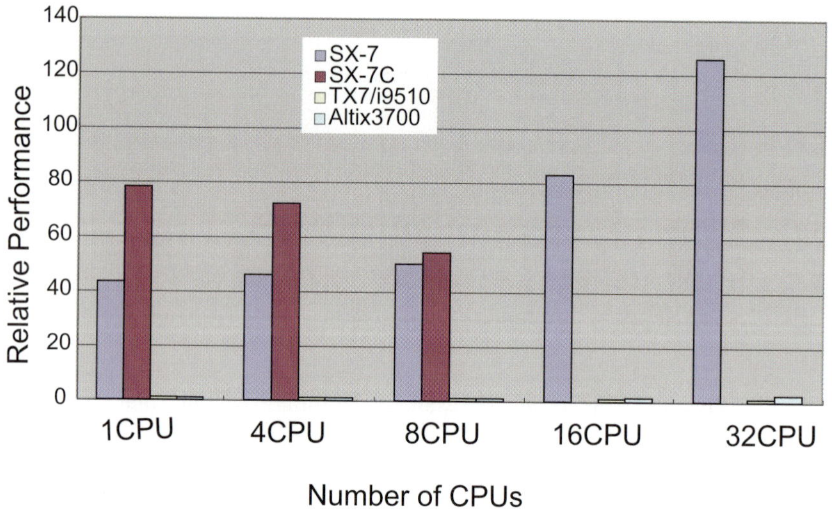

Fig. 22. Relative performance in EM-Scatter normalized by i9510

As Fig. 24 suggested, the scalability of TX7/i9510 stops when using 16 CPUs or more, although the Altix system scales well up to 32 CPUs. This is because of the difference in memory bandwidth per CPU between TX7 and Altix. The four CPUs of a TX7/i9510 cell share a 6.4 GB/s memory, and each node has up to 8 cell cards. Therefore, within 8 CPUs, each cell has only one CPU, and each CPU can exclusively use a 6.4 GB/s bus for memory reference. However, when using 16 or 32 CPUs, each cell has 2 or 4 CPUs, and the bandwidth available for each CPU is reduced to half or quarter of the peak bandwidth. This limitation seriously restrict the scalability of the

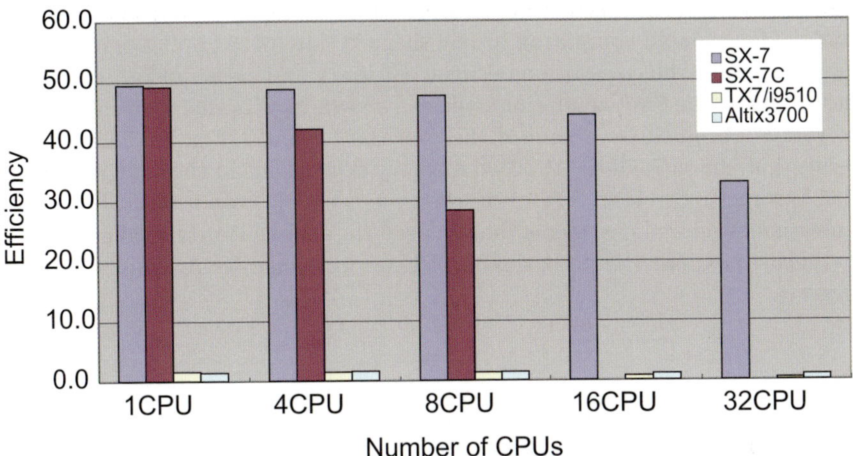

Fig. 23. Efficiency in execution of EM-Scatter

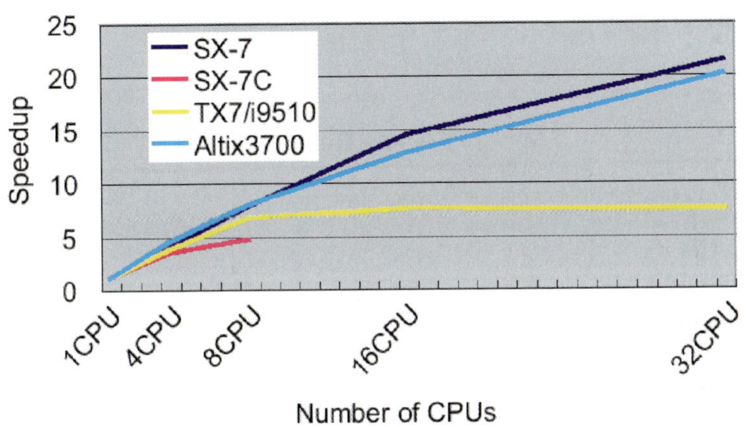

Fig. 24. Scalability in execution of EM-Scatter

TX7/i9510, resulting in a growing gap in performance SX-7 and TX7/i9510 in Fig. 22.

The four CPUs of a cell card of Altix also share the memory bus, however, two CPUs, not four CPUs share a 6.4 GB/s bus. Therefore, as each node has up to 16 cell cards (SGI calls a cell a fabric), even in the case of 32-CPU processing, each CPU can make a full use of a 6.4 GB/s memory bus and memory bus contentions are relaxed compared with TX7/i9510, resulting in the different behaver in scalability shown in Fig. 24. These results also suggest that even for scalar systems, at least one byte per flop rate is necessary to achieve the scalable computing in real memory-intensive applications.

We examine the correlation between efficiency and cache hit rates by using the results of the five simulation codes as shown in Figs. 25. Here, the cache hit rate is obtained by a sum of those of the L2 and L3 caches. This figure means that the ratio of the exposed memory access time in the total execution time becomes more than 50% even when the cache hit rate is 95%. Therefore, tuning codes on scalar systems that achieve the cache hit rate of almost 100% is crucial to exploit the potential of CPU performance, as Amdahl's low also suggests.

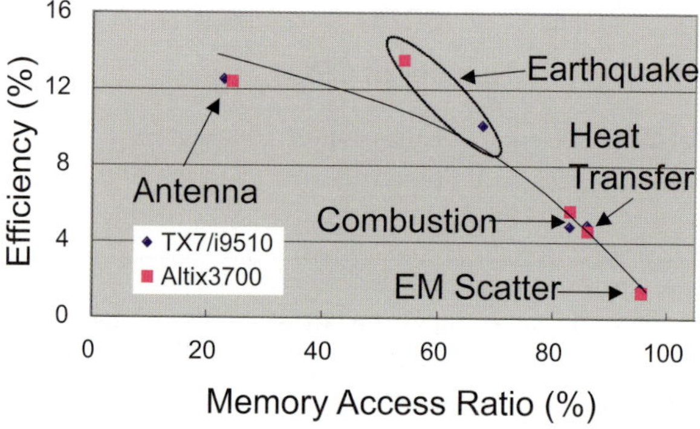

Fig. 25. Relationship between efficiency and cache hit rates of the scalar system

Implication of Memory Performance for Sustained System Performance on the Vector Systems

We examine how the memory bandwidths of the vector systems contribute to their overwhelming performance against the scalar systems in execution of the real simulation codes. Figure 26 shows the ratio of the sustained performance to the peak performance when limiting the memory bandwidth available for each CPU of SX-7, i.e., in the cases of the memory bandwidths of 8.83, 17.7 and 35.3 GB/s per CPU (4B/flop, 2B/flop and 1B/flop, respectively). As the figure clearly suggests, the sustained system performance seriously goes down as the memory bandwidth decreases. If the memory bandwidth is reduced to the level of scalar system, i.e., 1B/flop, the sustained performance is also reduced to the same level as the scalar systems, around 10% to 20%. These results mean that the excellent performance of the vector system are strongly supported by the excellent memory performance, and therefore, to keep the sustained performance higher, sufficient memory bandwidth for peak-flop/s is essential.

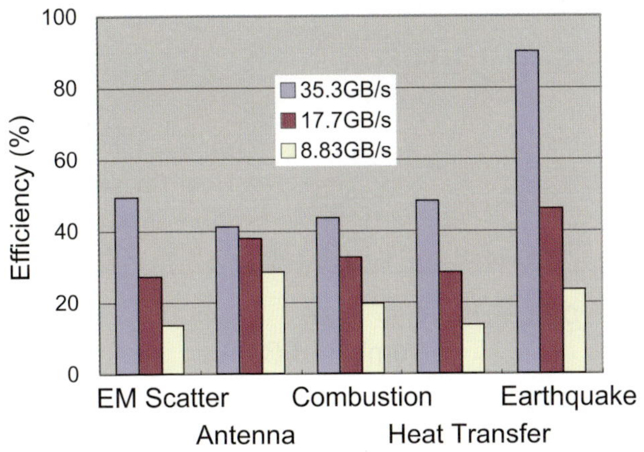

Fig. 26. Effect of memory bandwidths on SX-7 performance

We also examine the effect of memory banks of the vector systems on the sustained system performance. Here we calculate the minimum number of banks to hide the latency of memory for continuous memory access using load/store pipelines. The minimum numbers of banks per CPU is 356(=(simultaneous load/stores) × (bank busy time) / (clock cycle time) = 4 × 80[ns] / 0.9[ns]) for SX-7 and 512 (4 × 64[ns] / 0.5[ns]) for SX-7C, respectively. In the actual design of SX-7 and SX-7C, both systems have 512 banks per CPU. Therefore, SX-7 has a certain margin even when using full CPUs of a single node (32 CPUs for SX-7 and 8 CPUs for SX-7C).

Figure 27 shows the relationship between the number of banks and the efficiency in performance of SX-7. When the sufficient memory banks are provide to each CPU, for example 2K or more, the system keeps higher efficiency. However, when the number of banks per CPU go beyond 1K or less as the number of CPUs increases, there is a sharp drop in efficiency.

The bank shortage is more serious for SX-7C. Figure 28 shows the comparison of the efficiency of SX-7 and SX-7C when increasing the number of CPUs (decreasing the number of banks per CPU). When the number of CPUs reaches 8 in SX-7C, the number of banks also reaches the minimum number of banks (512 banks). In this situation, the efficiency of SX-7C (and SX-8 with the same architecture) becomes half of the SX-7, which means that the sustained system performance of the both systems is the same even though SX-7C (and SX-8) has twice the peak performance of SX-7. SX-7 also shows the performance degradation due to the lack of memory banks on 32 CPUs, but as the minimum number of banks per CPU for SX-7 is 356 and 512K banks per CPU are available even in 32-CPU processing, the reduction rate stays 15% to 25% in the efficiency of SX-7 performance. Consequently, enough memory banks, for example, at least 4 times as many as the minimum num-

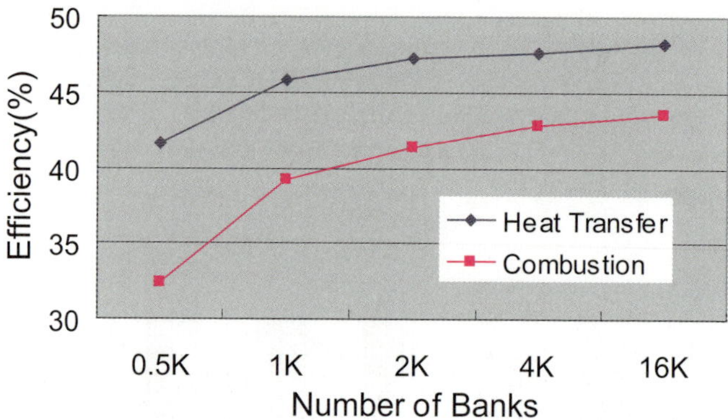

Fig. 27. Effect of the number of memory banks on SX-7 performance

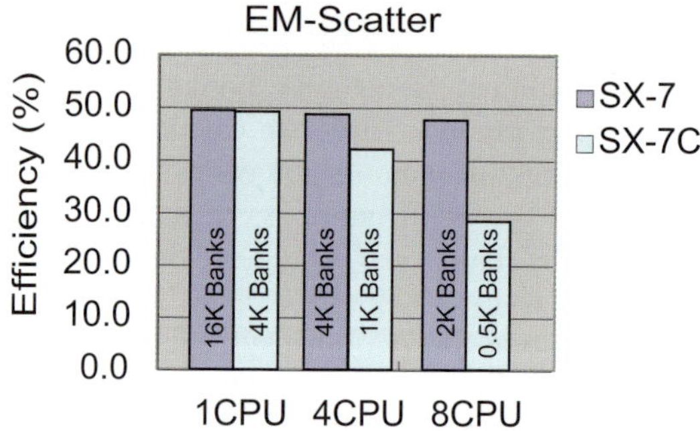

Fig. 28. Comparison of the effect of the number of memory banks on SX-7 and SX-7C performance

ber of banks, are also an important factor to keep higher sustained system performance for vector systems in execution of practical simulation codes.

4 Performance Evaluation of a Montecito Core

In the last section, we discuss the performance of the latest scalar system named TX7/i9610, which is equipped with Intel Montecito processors we have installed in April 2006. We think this is the first performance evaluation of the Montecito-based scalar system in the world.

4.1 TX7/i9610

Figure 29 shows the node architecture of TX7/i9610. Each cell card consists of four dual-core Montecito processors, and two processors (four cores) share a 8.5 GB/s memory bus. Each node of TX7/i9610 has up to 8 cell cards, and node performance reaches 410 Gflop/s (6.4 Gflop/s per Core). A Montecito core has an enlarged L3 cache of 12MB, and the FSB is also accelerated by 30% from that of TX7/i9510. This memory-related improvement is expected to contribute to an increase in sustained system performance even though the CPU clock frequency of 1.6 GHz is the same as that of TX7/i9510

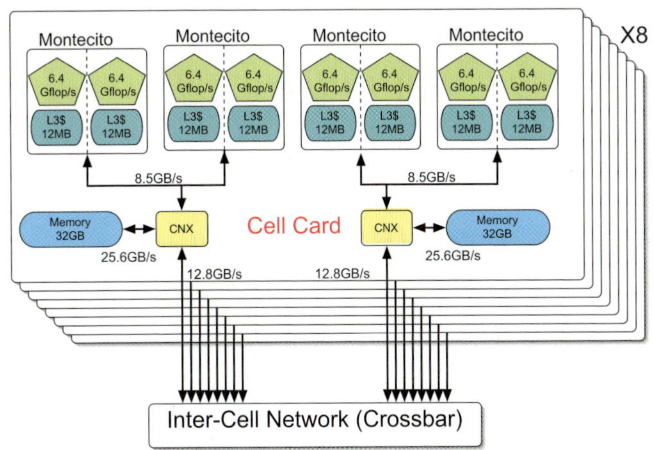

Fig. 29. The node architecture of TX7/i9610

4.2 Experimental Results and Discussion

Currently, as the i9610 is still under developing[1], we discuss the results of one Montecito Core evaluation. Figure 30 shows the performance of one Montecito core of TX7/i9610 normalized by i9510's performance discussed in the previous section. For comparison, the performance of Altix 3700 discussed in the previous section is also provided. In comparison between TX7/i9510 and TX7/i9610, increases in FSB bandwidth (from 400MHz to 533MHz) and L3 cache capacity (from 9MB to 12MB) contribute to the performance improvement of 15% to 86%, although their peak performances in flop/s are the same. In addition, the modification in the cache coherence protocol from the tag-based snoopy cache of i9510 to the directory-based one of i9610 also reduces the memory-related overhead even in the 1-CPU case. Therefore, these experimental results also suggest that memory-related tuning in the system architecture is very important to increase the sustained system performance.

---

[1] At the time of the 4th Teraflop workshop held on March 30-31, 2006

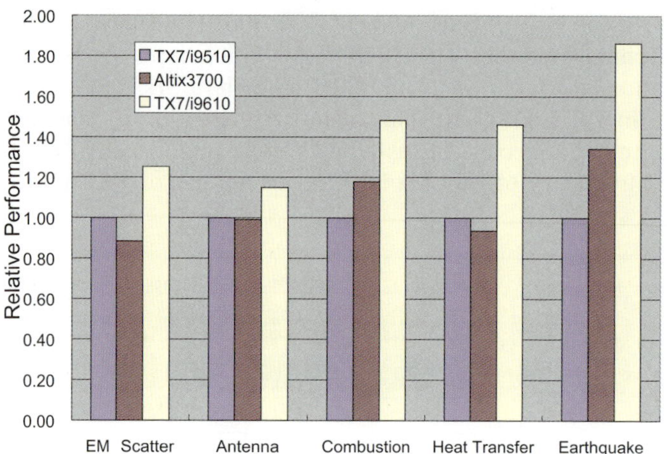

Fig. 30. Performance of single Montecito core

5 Conclusions

In this paper, we have presented the performance evaluation of modern vector-parallel and scalar parallel HEC systems. In the evaluation using the HPC challenge benchmarks, the vector systems can achieve quite impressive performance on the memory-related tests, although the number of CPUs of the vector systems is relatively small compared with the other scalar systems. On the other hand, the scalar systems with a large number of CPUs show the best performance on the global tests that can be increased as the number of CPUs increases. Because there is no universal solution in the HPC community, we have to find the best solution by considering application characteristics in terms of the memory reference behavior. For this purpose, the HPC challenge benchmark is very useful that evaluates HEC system from the viewpoint of the locality of memory reference of applications. In the applications in which their temporal/spatial locality is not so high, we think that an excellent balance of memory bandwidth per flop/s, 4 bytes/flop of the SX systems, is an important factor to realize highly efficient supercomputing. Of course, cache-friendly applications should be executed on the scalar-type HEC system from the cost/performance perspective.

In the evaluation using the leading practical application codes in the fields of electro-magnetic analysis, CFD/heat transfer analysis, and earthquake analysis, we have presented that memory-performance is a very important factor to increase the sustained system performance. The vector systems with 10 times faster memory bandwidth show overwhelming performance compared with the scalar systems. We have also pointed out that to keep the higher sustained performance, a larger number of memory banks and a higher bandwidth-flop/s rate are crucial factors for the vector systems. Therefore, if the vector systems would lose the advantage regarding memory performance against the scalar systems in the future design, they would also lose the certain position in the HPC community.

Japan's Ministry of Education, Culture, Sports, Science and Technology (MEXT) just starts a new supercomputer R&D project named *Development and Applications of Advanced, High-Performance, Supercomputer Project.* In this project, a national-leadership supercomputer, which has a sustained speed of 10 peta flop/s, will be designed and developed, and the system is supposed to be available in 2010. The total investment is estimated to be 1 billion US dollar for the next seven years. We believe no matter what kind of architectures will be employed in the system, how high-performance memory systems can be incorporated into the system is a key to make a success of the project, to avoid spending a huge money only for marking a top-1 score on TOP500 ranking. For the vector architecture, whether the golden rule of 4 bytes/flop can be kept in the future systems or not might decide the future of the vector architecture. Lowering power and energy consumption of future HEC systems is serious design factor. We are doing R&D of the next generation, low-power, highly efficient vector architectures in collaboration with NEC.

Acknowledgments

This work has been done in collaboration between Tohoku University and NEC, and many colleagues contribute to this project. In particular, we would like to acknowledge Koki Okabe, Hiroyuki Takizawa of Tohoku University and Akihiko Musa, Tatsunobu Kokubo, Takashi Soga, and Naoyuki Sogo of NEC. We would also like to thank Professors Motoyuki Sato, Akira Hasegawa, Goro Masuya, Terukazu Ota, and Kunio Sawaya of Tohoku University for providing the codes for the experiments.

References

1. K.Ariyoshi et al. Spatial variation in propagation speed of postseismic slip on the subducting plate boundary. *Proceedings of 2nd Water Dynamics*, 35, 2004.
2. K.Kitagawa et al. A Hardware Overview of SX-6 and SX-7 Supercomputer. *NEC Research & Development*, 44(1):2–7, 2003.
3. K.S.Kunz and R.J.Luebbers. *The Finite Difference Time Domain Method for Electromagnetics.* CRC Press, 1993.
4. K.Tsuboi and G.Masuya. Direct Numerical Simulations for Instabilities of Remixed Planar Flames. *Proceedings of The Fourth Asia-Pacific Conference on Combustion*, 2003.
5. L.Oliker et al. Scientific Computations on Modern Parallel Vector Systems. *Proceedings of SC2004*, 2004.
6. L.Oliker et al. Leading Computational Methods on Scalar and Vector HEC Platforms. *Proceedings of SC2005*, 2005.

7. MEXT HPC Task Force. (in Japanese) 2nd Report on Next Generation High-Performmace Computing Systems and their Applications in Japan. 2005.
8. M.Nakajima et al. Numerical Simulation of Three-Dimensional Separated Flow and Heat Transfer around Staggerd Suerface-Mounted Rectangular Blocks in a Channel. *Numerical Heat Transfer*, 47(Part A):691–708, 2005.
9. P.Luszczek, J.Dongarra, D.Koester, R.Rabenseifner, B.Lucas, J.Kepner, J.McCalpin, D.Bailey, and D.Takahashi. Introduction to the HPC Challenge Benchmark Suite. *http://icl.cs.utk.edu/hpcc/ubs/index.html*, 2005.
10. The High-End Computing Revitalization Task Force. Federal Plan for High-End Computing. *Technical report*, 2004.
11. T.Kobayashi et al. FDTD simulation on array antenna SAR-GPR for land mine detection. *Proceedings of 1st International Symposium on Systems and Human Science*, pages 279–283, 2003.
12. Top 500 Supercomputer Sites. http://www.spec.org/.
13. T.Senta et al. Itanium2 32-way Server System Architecture. *NEC Research & Development*, 44:8–12, 2003.
14. Y.Takagi et al. Study of High Gain and Broadband Antipodal Fermi Antenna with Corrugation. *Proceedings of 2004 International Symposium on Antennas and Propagation*, 1:69–72, 2004.

Recent Performance Results of the Lattice Boltzmann Method

Peter Lammers and Uwe Küster

Höchstleistungsrechenzentrum Stuttgart,
University of Stuttgart,
Nobelstrasse 19, 70569 Stuttgart, Germany
lammers|kuester@hlrs.de

1 Abstract

In this paper we present performance results for the lattice Boltzmann method on three different architectures. The benchmarked architectures include an IBM p575, a SGI Altix 3700 (LRZ Munich) and a NEC SX-8 (HLRS Stuttgart). The application used is the well known lattice Boltzmann solver $\mathcal{BEST}$. Furthermore we use a modified lattice Boltzmann solver using a boundary-fitted cartesian mesh to compare the performance of indirect addressing on the NEC SX-8 with the predecessor model SX-6+.

2 Introduction

In the past decade the lattice Boltzmann method (LBM) [6,7,9] has been established as an alternative for the numerical simulation of (time-dependent) incompressible flows. One major reason for the success of LBM is that the simplicity of its core algorithm allows both easy adaption to complex application scenarios as well as extension to additional physical or chemical effects. Since LBM is a direct method, the use of extensive computer resources is often mandatory. Thus, LBM has attracted a lot of attention in the High Performance Computing community [1,3,5]. Here we use the characteristics of the LBM to gain a deep understanding of the performance characteristics of modern computer architectures which might be useful to develop appropriate optimization strategies that can also be applied to other applications. We evaluate the performance for different pairs of processor numbers and grid points. The results can be either viewed in an Amdahl or in a Gustafson way. This kind of analysis can easily be performed because of the simplicity of the method. Especially the feature that a traditional grid generation is unnecessary as the underlying grid for LBM is a cartesian equidistant mesh. Geometries are handled by the marker-and-cell approach where solid nodes are blocked out in the grid. Furthermore the domain decomposition and load balancing can be done in the solver itself.

In this report we discuss LBM performance in the mentioned way on three tailored HPC systems including an IBM p575, a SGI Altix 3700Bx2 and a NEC SX8 vector system. For those studies we use the LBM solver $\mathcal{BEST}$ (Boltzmann Equation Solver Tool), which is written in FORTRAN90 and parallelized with MPI using domain decomposition. As a test case we run simulations of flow in a long channel with square cross-section, which is a typical application in turbulence research.

3 Lattice Boltzmann method

Nowadays, the lattice Boltzmann method is applied in a wide range of scientific and engineering disciplines including chemical and process engineering, bio and environmental processes and medical applications. Especially it is preferred when highly complex geometries are involved. The first author of the article is intensively using the lattice Boltzmann method for simulations of flow control in wall bounded turbulence, turbulent drag reduction and control of transition [2] (see Fig. 1). Among chemical engineering the flow solver $\mathcal{BEST}$ was developed at the Lehrstuhl für Strömungsmechanik (LSTM), Universität Erlangen-Nürnberg for this purpose. Here, the $\mathcal{BEST}$ solver is used for most of the benchmarks whose results are shown in this paper. The benchmark itself (a plane channel) is motivated by wall bounded turbulence flows.

The lattice Boltzmann method consist of a discretized kinetic equation for the one particle distribution function f,

$$f_\alpha(\mathbf{x} + \mathbf{c}_\alpha, t + 1) = f_\alpha(\mathbf{x}, t) - \frac{1}{\tau} \left\{ f_\alpha(\mathbf{x}, t) - f_\alpha^{eq}(\mathbf{x}, t) \right\}, \tag{1}$$

for which an appropriate phase velocities lattice need to be specified. In $\mathcal{BEST}$, the 3D spatial cartesian grid is coupled to the D3Q19 streaming

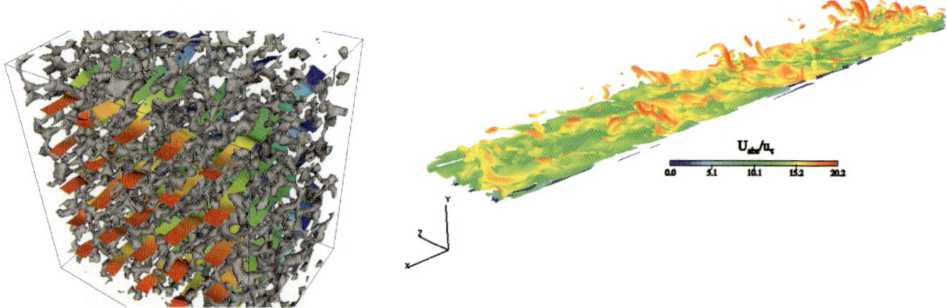

Fig. 1. Lattice Boltzmann simulation of a flow through a highly porous media (left) and a direct numerical simulation of wall bounded turbulence (right)

lattice [6] with 19 mesoscopic variables $\mathbf{c}_\alpha$. τ is a relaxation parameter that determines the rate at which the particle interaction drives f to the equilibrium state f^{eq}. f^{eq} is appropriately chosen to satisfy the Navier-Stokes equation and depends on the hydrodynamic quantities like density, pressure and the velocity fields which are calculated from the moments of f.

In the benchmark codes, the algorithm is implemented just as given by eq. (1). That means, a collision step (r.h.s) in which basically the equilibrium state function f^{eq} is calculated, is followed by a propagation step (l.h.s) in which the mesoscopic variables f_α are streamed along the lattice directions $\mathbf{c}_\alpha$. In order to reduce the memory traffic both steps are combined. As the streaming step implicates a data dependency, it cannot be vectorized in a straightforward way. In the implementation this is by-passed by using two arrays of f for time step t and $t + 1$ which are used in an alternating way. For cache based machines the inner loop of the main kernel is broken up in pieces. Especially, the propagation step is done for pairs of the densities f_α.

For parallelization the overall spatial grid is block-wise distributed to the processors in $\mathcal{BEST}$. At the boundaries, the propagating densities are first copied to a ghost layer and then sent to the adjacent processor block. The data exchange is realized by `MPI_Sendrecv` which is at least the optimal choice on the SX network for this communication pattern. For more details of the implementation see [2] and [8]

4 Performance evaluation

In each of the following plots the CPU efficiency depending on the domain size per process is plotted. Several grid sizes are chosen ranging from 500 to 25 mio. grid points. Especially for vector architecture this means that even for the smallest case the vector length which is identical with the domain size is sufficient. The right axis correlates the given efficiency to the theoretical peak performance of one CPU. In case of the vector system the values are based on the vector peak performance.

Additionally, in the same plot the curves for parallel runs are included. For a fixed size per process the values on different curves show the weak-scaling scenario. For this kind of presentation, ideal scalability means that all curves would collapse into one.

In some plots also performance numbers in MLUPS (**Mega Lattice Site Updates per Second**) are given. This is a handy unit for measuring the performance of LBM and gives the time to calculate one time step for a given grid.

4.1 Performance on the IBM p575

The IBM Power5 (Performance optimized with enhanced RISC) processor is an enhancement of the 64-bit Power4 architecture. The two multiply-add floating point units share a three-bank L2 Cache of 1.92 MB (3 x 640 KB caches

with independent buses, 10-way set associative). With the clock frequencies of
1.9 GHz the theoretical peak performance is 7.6 GFlops/s. An important difference to the Power4 is that a L3 cache and a memory controller are on-chip
now. A large 36 MB L3 cache operate at half the CPU speed. This should
overcome the limited memory bandwidth observed in the Power4 servers.

We present here only the results for a 8-way compute node with 32 GB
memory. The measurements on the Power5 were made by IBM [4]. At the
computing center of the Max Planck society in Garching, an installation with
86 node is available but the nodes are connected by a Gbit ethernet, which
would lead to poor inter-node performance. On the other hand it is known
that the IBM federation switch scales quite nice for the examined application
and hence is an obvious choice.

The results of the single and parallel performance of $\mathcal{BEST}$ are shown in
Fig. 2. The single processor performance is always given by the dotted black
lines. The achievable performance per process of the Power5 CPU (Fig. 2) is in
the range between 16.5 % and a maximum of 25 % efficiency, which correspond
to 1.25 and 1.9 GFlop/s respectively. Significant outliers due to cache effects
can be observed for inner loop lengths of 16, 24 and 32. The results for 8 CPUs
are represented by the red curve. Here the important observation is that in
contrast to the Power4 the p575 provides sufficient memory bandwidth inside
one node now.

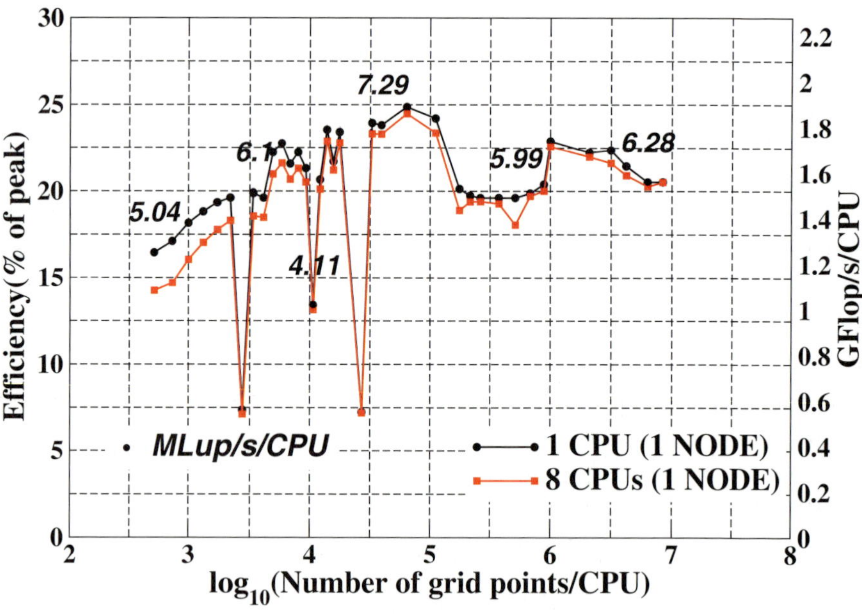

Fig. 2. Efficiency, GFlop/s and MLup/s of the lattice Boltzmann solver
$\mathcal{BEST}$ depending on the domain size and the number of processors for up to 8
CPUs of an IBM p575

4.2 Performance on the SGI Altix

An important competitor for the IBM server as well as for the vector system is the SGI Altix architecture which is based on the Intel Itanium 2 processor. This CPU has a superscalar 64-bit architecture providing two multiply-add units and uses the Explicitly Parallel Instruction Computing (EPIC) paradigm. Contrary to traditional scalar processors, there is no out-of-order execution. Instead, compilers are required to identify and exploit instruction level parallelism. Today clock frequencies of up to 1.6 GHz and on-chip caches with up to 9 MBytes are available. The basic building block of the Altix is a 2-way SMP node offering 6.4 GByte/s memory bandwidth to both CPUs, i.e. a balance of 0.06 Word/Flop per CPU. The SGI Altix3700Bx2 (SGI Altix3700) architecture as used for the $\mathcal{BEST}$ application is based on the NUMALink4 interconnect, which provides up to 3.2 GByte/s bidirectional interconnect bandwidth between any two nodes and latencies as low as 2 microseconds. The NUMALink technology allows to build up large powerful shared memory nodes with up to 512 CPUs running a single Linux OS.

The Itanium2 (see Fig. 3) achieves a maximum of 36 % efficiency corresponding to 2.3 GFlop/s. Performance drops significantly when the problem size exceeds the cache size. Further increasing the problem size, compiler-

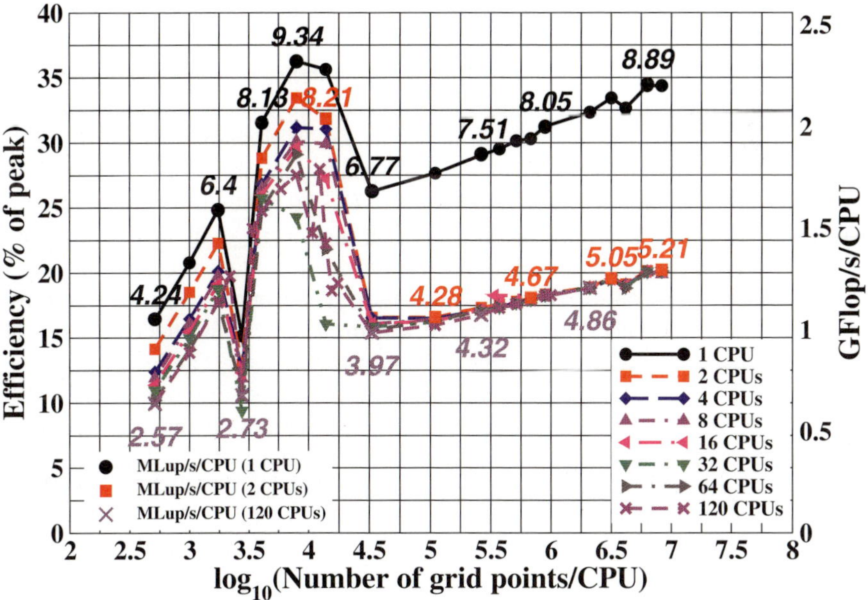

Fig. 3. Efficiency, GFlop/s and MLup/s of the lattice Boltzmann solver $\mathcal{BEST}$ depending on the domain size and the number of processors for up to 120 CPUs of a SGI Altix3700Bx2

generated prefetching takes over and leads to gradual improvement up to a final level of 2.2 GFlop/s.

In the SGI Altix two processors have to share one memory connection. Consequently when using two processors single CPU performance drops to 1.29 GFlop/s. Prefetching is still important but not as efficient as for one processor. Furthermore, in Fig. 3 weak scaling results for up to 120 Itanium 2 CPUs are given revealing that the scaling inside a node is very satisfactory.

From a programmer's view, the NEC SX-8 is a traditional vector processor with 4-track vector pipes running at 2 GHz. One multiply and one add instruction per cycle can be sustained by the arithmetic pipes, delivering a theoretical peak performance of 16 GFlop/s. The memory bandwidth of 64 GByte/s allows for one load or store per multiply-add instruction, providing a balance of 0.5 Word/Flop. The processor has 64 vector registers, each holding 256 64-bit words. Basic changes compared to its predecessor systems are a separate hardware square root/divide unit and a "memory cache" which lifts stride-2 memory access patterns to the same performance as contiguous memory access. An SMP node comprises of eight processors and provides a total memory bandwidth of 512 GByte/s. The SX-8 nodes are networked by an interconnect called IXS, providing a bidirectional bandwidth of 16 GByte/s and a latency of about 5 microseconds.

4.3 Performance on the NEC SX

Results for the SX-8 are shown in Fig. 4. In contrast to the systems seen so far, the performance of the vector system increases with increasing grid size and saturates at an efficiency close to 75 %, i.e. at a single processor application performance of 11.9 GFlop/s. This is partly an effect of the increasing vector length and on the other hand reflects the fact that the computational part dominates the communication part with increasing domain size. But the network is far from perfect scalability as can be seen from the results for up to 576 CPUs. We believe that parasitic non synchronous operating system processes show influences on aggregated waiting times. This effect becomes of course smaller by increasing the per process problem size. The load balancing itself is ideal for this benchmark case and can not cause any problems.

Overall, we see for the lattice Boltzmann application on the 576 processor NEC SX-8 system a maximum sustained performance of 5.7 TFlop/s. The same sustained performance level would require at least 6400 Itanium2 CPUs on an SGI Altix3700.

Finally, we would like to focus in this context on the performance of indirect array addressing on the SX-8. In $\mathcal{BEST}$ the calculation is done on a block structured regular grid. But for highly complex geometries it is of course more appropriate to store only the cells inside the flow domain. For this, an index is needed to select the fluid cells and the information about adjacent nodes has to be stored in index lists.

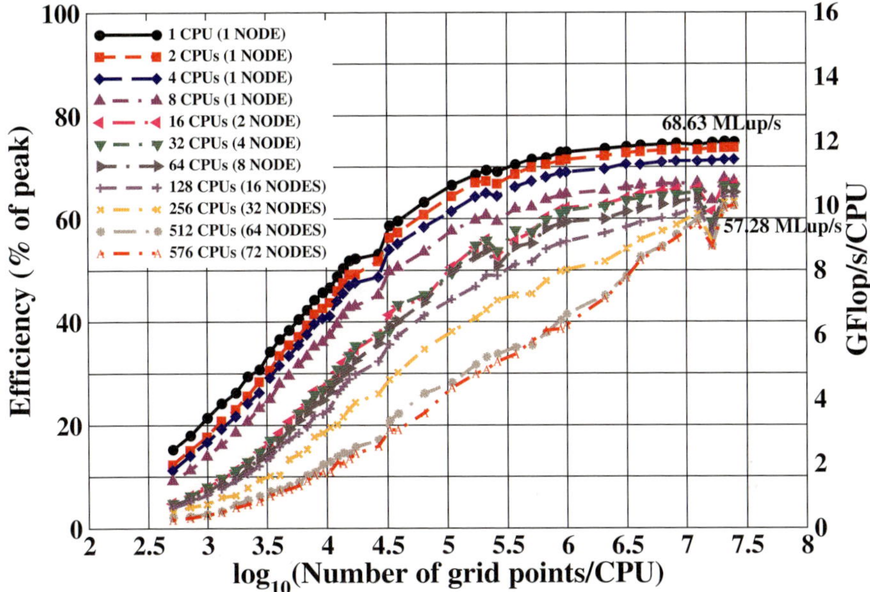

Fig. 4. Efficiency and GFlop/s of the lattice Boltzmann solver $\mathcal{BEST}$ depending on the domain size and the number of processors for up to 72 nodes (576 CPUs) of a NEC SX-8

In Fig. 5, performance on the SX-6+ and SX-8 of such an implementation of the lattice Boltzmann method is plotted. The NEC SX-6+ system implements the same processor technology as used in the Earth Simulator but runs at a clock speed of 565 MHz instead of 500 MHz. In contrast to the NEC SX-8 this vector processor generation is still equipped with two 8-track vector pipelines allowing for a peak performance of 9.04 GFlop/s per CPU for the NEC SX-6+ system. Note that the balance between main memory bandwidth and peak performance is the same as for the SX-8 (0.5 Word/Flop) both for the single processor and the 8-way SMP node.

The code used for the measurement is a collaborative development between RRZE, Institut für Computeranwendungen im Bauingenieurwesen (CAB), Universität Braunschweig, Department of Computational Science, University of Amsterdam, Department of Medical Physics and Clinical Engineering, University of Sheffield, NEC CCRL and HLRS. The test cases are identical with the cases used for the previous plots.

Till now only a serial code version is available. This explains the slightly better performance for small grids. In the results shown so far, the communication dominates the calculation for small grids. The single CPU efficiency on SX-6+ is around 60%. On the other hand the efficiency of the SX-8 is still 50%, anout 10% lower than on the SX-6+. The reason is the higher memory latency of the recent system. Therefore it is more important on the SX-8 to

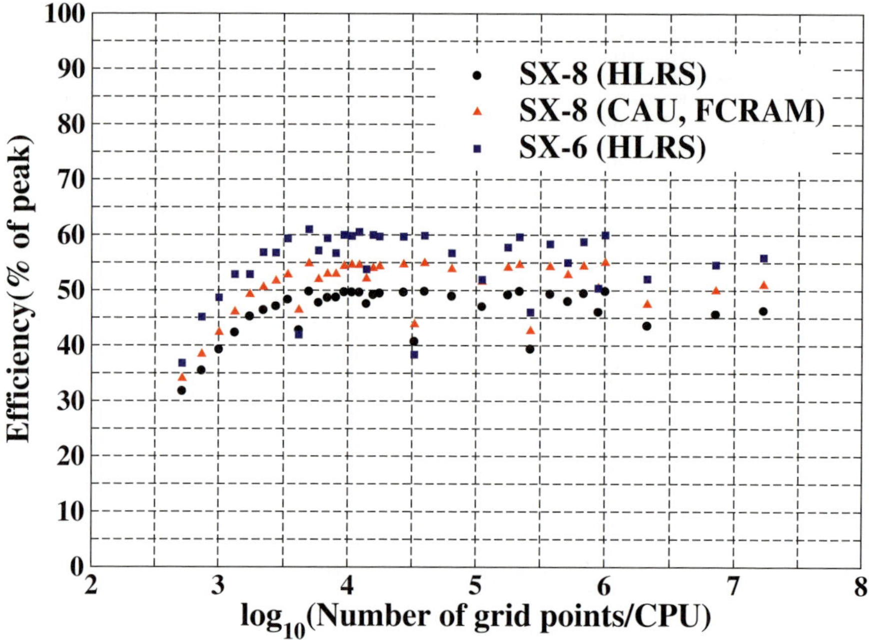

Fig. 5. Performance (efficiency) of an unstructured implementation of the lattice Boltzmann kernel on the SX-8 in comparison to the SX-6

prefetch the vector gather/scatter instructions for the indirect load/stores and hide the latency by other instructions. This can typically be done by unrolling simple loops. The present loop is complex enough for hiding these latencies partially.

The SX-8 is also available with fast cycle memory (FCRAM). With FCRAM the memory per node is limited to 64 GB, instead of a maximum of 128 GB with DDR2-SDRAM. At the computing center of the Christian-Albrechts-University Kiel they are operating a system of this kind with 5 8-cpu compute nodes. Figure 5 includes also the performance results for a run on this machine in Kiel. With FCRAM instead of DDR2-SDRAM the effect is not so pronounced however still about 5% lower than the efficiency on SX-6+.

Outlook

From the results for the SGI Altix 3700 we expect the new HLRB II system installed at the LRZ in the middle of 2006 to be a real competitor to the NEC system at HLRS. It will be a SGI Altix 4700 with 4096 cores. On this system the performance drop between one and two CPUs should not occur because of the better memory connection.

It would also be interesting to have results for the IBM Blue Gene which is dominating the top500 list although this architecture is not aiming at flow simulations. At the NIC in Jülich an installation with more than 40 TFlops theoretical peak performance is available.

References

1. J. Carter, M. Soe, Leonid Oliker, Y. Tsuda, G. Vahala, L. Vahala, and A. Macnab. Magnetohydrodynamic Turbulence Simulations on the Earth Simulator Using the lattice Boltzmann Method. Supercomputing 2005, 2005.
2. P. Lammers. *Direkte numerische Simulationen wandgebundener Strömungen kleiner Reynoldszahlen mit dem lattice Boltzmann Verfahren.* Dissertation, Universität Erlangen–Nürnberg, 2005.
3. F. Massaioli and G. Amati. Achieving high performance in a LBM code using OpenMP. In *EWOMP'02*, Roma, Italy, 2002.
4. Jakob Pichlmeier. IBM, private communication, 2005.
5. T. Pohl, N. Thürey, F. Deserno, U. Rüde, P. Lammers, G. Wellein, and T. Zeiser. Performance Evaluation of Parallel Large-Scale Lattice Boltzmann Applications on Three Supercomputing Architectures. In *Supercomputing 2004*, Nov. 2004. Supercomputing Conference 04.
6. Y. H. Qian, D. d'Humières, and P. Lallemand. Lattice BGK models for Navier-Stokes equation. *Europhys. Lett.*, 17(6):479–484, January 1992.
7. Sauro Succi. *The Lattice Boltzmann Equation – For Fluid Dynamics and Beyond.* Clarendon Press, 2001.
8. G. Wellein, T. Zeiser, G. Hager, and S. Donath. On the single processor performance of simple lattice boltzmann kernels. *Computers & Fluids*, 2005. In Press, Corrected Proof, Available online 20 December 2005.
9. Dieter A. Wolf-Gladrow. Lattice-Gas Cellular Automata and Lattice Boltzmann Models. 1725, 2000.

Linear Iterative Solver for NEC Parallel Vector Systems

Sunil R. Tiyyagura and Uwe Küster

High Performance Computing Center Stuttgart,
University of Stuttgart,
Nobelstrasse 19, 70569 Stuttgart, Germany
sunil|kuester@hlrs.de

Summary. This paper addresses the performance issues of linear iterative solver on vector machines. Preconditioned iterative methods are very popular in solving sparse linear systems arising from discretizing PDEs using Finite Element, Finite Difference and Finite Volume methods. The performance of such simulations heavily depends on the performance of the sparse iterative solver, as major portion of the time is spent here. First, the performance of widely used public domain solvers is analysed on NEC SX-8. Then, a newly developed parallel sparse iterative solver targeting vector machines (Block-based Linear Iterative Solver – BLIS) is introduced. Finally, the performance gain by the presented modifications is demonstrated.

Keywords: Linear iterative solver, Indirect memory addressing, Vector architecture, Sparse matrix vector product.

1 Introduction

Recent trend in supercomputing is directed towards adding thousands of scalar processors connected with innovative interconnects. This is naturally a setback to vector computing in general and development of new vectorizable algorithms in particular. The major concern with this developing trend is the scalability of communication intensive applications which need global synchronization at many points, for example a conjugate gradient solver. With the increase of processor count beyond a certain point (typically 1000), the scalability of such applications may be seriously inhibited. Vector architectures seem to be a better alternative for such applications by providing a very powerful processor (3-4 times more powerful than a commercial scalar processor). These architectures are typically clusters of powerful SMP (Symmetric Multiprocessing) nodes that reduces the overhead for synchronization as each node can process a huge amount of computation.

Another challenge facing computational scientists today is the rapidly increasing gap between sustained and peak performance of the high performance computing architectures. Even after spending considerable time on tuning applications to a particular architecture, this gap is an ever existing problem. A comprehensive study of this problem both on scalar and vector machines can be found in [1, 2]. The sustained computation to communication ratio and the computation to memory bandwidth ratio are much better for vector architectures when compared to clusters of commodity processors [3]. The success of the Earth Simulator project [4] and the ongoing Teraflop Workbench project [4] emphasizes the need to look towards vector computing as a future alternative for certain class of applications.

In this paper we focus on the performance of linear iterative solver on vector machines. Firstly, performance of AZTEC [6], a widely used public domain solver is analysed on NEC SX-8. The reasons for the dismal performance of public domain solvers on vector systems are then stated. A newly developed sparse iterative solver is introduced and reasons for its superior performance elaborated. All the performance measurements of the iterative solver are done with Finite Element (FE) applications using the research finite element program *Computer Aided Research Analysis Tool* (CCARAT), that is jointly developed and maintained at the Institute of Structural Mechanics of the University of Stuttgart and the Chair of Computational Mechanics at the Technical University of Munich. The research code CCARAT is a multipurpose finite element program covering a wide range of applications in computational mechanics, like e.g. multi-field and multi-scale problems, structural and fluid dynamics, shape and topology optimization, material modeling and finite element technology. The code is parallelized using MPI and runs on a variety of platforms, on single processor systems as well as on clusters.

In recognition of the opportunities in the area of vector computing, the High Performance Computing Center Stuttgart (HLRS) and NEC are jointly working on a cooperation project "Teraflop Workbench", whose main goal is to achieve sustained teraflop performance for a wide range of scientific and industrial applications. CCARAT is one of the workbench projects for simulating Fluid-Structure Interaction. The major time consuming portions of a finite element simulation are calculating the local element contributions to the globally assembled matrix and solving the assembled global system of equations. In Sect. 2 of this paper, we take a close look at the performance of public domain solvers on vector machines. In Sect. 3, the design features of the newly developed iterative solver targeting vector architecture is explained. A brief introduction to features of a typical vector architecture follows.

1.1 Vector Processor

Vector processors like NEC SX-8 use a very different architectural approach than scalar processors. Vectorization exploits regularities in the computational structure to accelerate uniform operations on independent data sets. Vector arithmetic instructions are composed of identical operations on elements of vector operands located in vector registers.

For non-vectorizable instructions the NEC SX-8 processor also contains a cache-based superscalar unit. Since the vector unit is by far more powerful than the scalar unit, it is crucial to achieve high vector operation ratios, either via compiler discovery or explicitly through code and data (re-)organization. The vector unit has a clock frequency of 2 GHz and provides a peak vector performance of 16 GFlop/s (4 add and 4 multiply pipes working at 2 GHz). The total peak performance of the processor is 22 GFlop/s (including divide/sqrt unit and scalar unit). Table 1 gives an overview of the different processor units.

Table 1. NEC SX-8 processor units

Unit	No. of results per cycle	Peak (GFlop/s)
Add	4	8
Multiply	4	8
Divide/sqrt	2	4
Scalar		2
		Total = 22

2 Public Domain Solvers

Here, we look into the performance of public domain sparse iterative solvers on vector machines. As much as 90% of the time in a very large scale simulation can be spent in the solver, specially if the problem to be solved is ill-conditioned. While the time taken for calculating and assembling of the global matrices (linear system) scales linearly with the size of the problem, often the time in the sparse solver does not. Major reasons being the conditioning of the problem and the kind of preconditioning needed for a successful solution.

Most public domain solvers like AZTEC [6], PETSc, Trilinos [7], etc. do not perform on vector architecture as well as they do on superscalar architectures. The main reason being their design considerations, that primarily target superscalar architectures. This is a direct consequence of the development in supercomputing that was directed towards clusters of commodity processors. This resulted to a huge gap in the development of software for vector machines. The recent success of vector architecture lead to some big installations which are helping to bridge this gap. The design of most of the available solvers effects the following performance critical features of vector systems.

2.1 Average Vector Length

This is an important metric that has a huge effect on vector performance. In sparse linear algebra, the matrix object is sparse where as the vectors are still dense. So, any operations involving only vectors, like the dot product, result in a good average vector length as the innermost vectorized loop runs over long vectors.

This is a problem for operations involving the sparse matrix object. Sparse matrix vector product (MVP) is a key kernel in a krylov subspace based iterative solver. The data structure used to store the sparse matrix plays an important role in the performance of such operators. Row based data structures used in the above mentioned solvers result in low average vector length which is further problem dependent. Though, they are an easy and natural way of representing a sparse matrix object, this hinders performance of one of the critical operators on vector machines.

A well known solution to this problem is to use pseudo diagonal data structure to store the sparse matrix [8]. A detailed overview of different sparse matrix formats and corresponding format conversions can be found in [9]. Notably, Trilinos provides a vector version of the sparse MVP operator to be used on vector machines. This ensures that the sparse matrix is converted from the native row format to a pseudo diagonal format before performing any operation that uses it. This pseudo diagonal sparse matrix object would need extra memory which could be a drawback if the problem size is already too large.

We tried to introduced similar functionality into AZTEC. Figure 1 shows the single CPU performance in Sparse MVP on the NEC SX-8 with the native row based (Modified Sparse Row – MSR) and the introduced pseudo diagonal based (JAgged Diagonal – JAD) matrix storage formats. The low performance for row based format (MSR) shown here can also be seen in the CG scientific kernel of the NAS parallel class B benchmarks where a performance of 470 MFlop/s was measured on NEC SX-6 [2]. Using vector registers to reduce memory operations for loading and storing the result vector further improves the performance of JAD based sparse MVP to 2.2 GFlop/s. By introducing some other optimization techniques, a maximum performance of 20% vector peak can be reached for sparse MVP on NEC SX-8 [10].

2.2 Indirect Memory Addressing

The performance of sparse MVP on vector as well as on superscalar architectures is not limited by memory bandwidth, but by latencies. Due to the sparse storage, the vector to be multiplied in a sparse MVP is accessed randomly (non-strided access). Hence, the performance of this operation completely depends on the implementation and performance of the "vector gather" assembly instruction on vector machines. Though the memory bandwidth and byte/flop ratios of a vector architecture are in general far more superior than any superscalar architecture, superscalar architectures have the advantage of cache re-usage for this operation. But, the cost of accessing the main memory is so high that without special optimizations [11], sparse MVP performs at around 5% peak on superscalar machines.

It is interesting in this context to look into architectures which combine the advantages of vector pipelining with memory caching, like the CRAY X1. Sparse MVP with pseudo diagonal format on this machine performs at about 14% peak [12]. This is similar to the performance achieved on the NEC SX-8. An upper bound of around 30% peak is estimated for most sparse matrix computations on the CRAY X1.

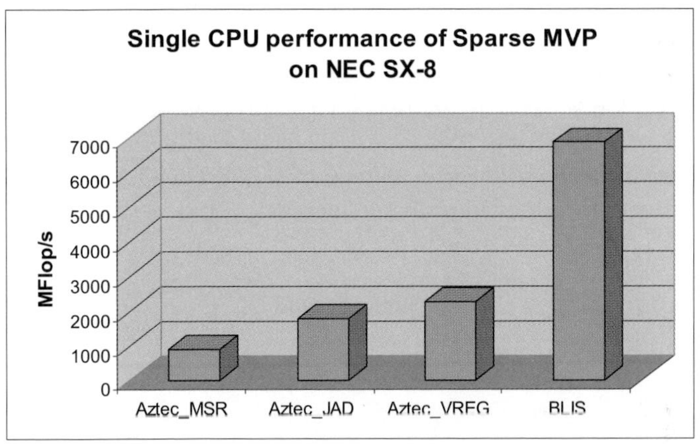

Fig. 1. Single CPU performance of Sparse MVP on NEC SX-8

3 Block-based Linear Iterative Solver – BLIS

In the sparse MVP kernel discussed so far, the major hurdle to performance is not memory bandwidth but the latencies involved due to indirect memory addressing. Block based computations exploit the fact that many FE problems typically have more than one physical variable to be solved per grid point. Thus, small blocks can be formed by grouping the equations at each grid point. Operating on such dense blocks considerably reduces the amount of indirect addressing required for sparse MVP [10]. This improves the performance of the key kernel dramatically on vector machines [13] and also remarkably on superscalar architectures [14, 15]. BLIS uses this approach primarily to overcome the penalty incurred due to indirect memory access.

3.1 Available Functionality

Presently, BLIS is working with finite element applications that have 4 unknowns to be solved per grid point. It is to be noted that the vectorizing loops include computation of whole blocks and hence lead to memory bank conflicts for even sized blocks. The performance degradation due to bank conflicts can be overcome by array padding and is included in BLIS. JAD sparse storage format is used to store the dense blocks. This assures sufficient average vector length for operations done using the sparse matrix object (Preconditioning, Sparse MVP). The single CPU performance of sparse MVP, Fig. 1, with a matrix consisting of 4x4 dense blocks is around 6.9 GFlop/s (about 40% vector peak) on the NEC SX-8.

The scaling behaviour of the solver on NEC SX-8 was tested for fluid applications that need to solve for 3 velocities and 1 pressure value at each discrete point on the grid using stabilized 3D fluid elements in CCARAT. Figure 2 plots strong scaling of BLIS upto 64 processors using BiCGSTAB algorithm on the NEC SX-8. Block Jacobi preconditioning was used for all the problems. It can be deduced from the plots that a problem size of about 10k finite elements per processor is needed to achieve a per processor performance of above 4 GFlop/s. This is far beyond the problem size needed to achieve good average vector length. The major hurdle in scaling is the need for global synchronization needed at various steps of the BiCGSTAB algorithm. This overhead is further amplified with the increase in processor count.

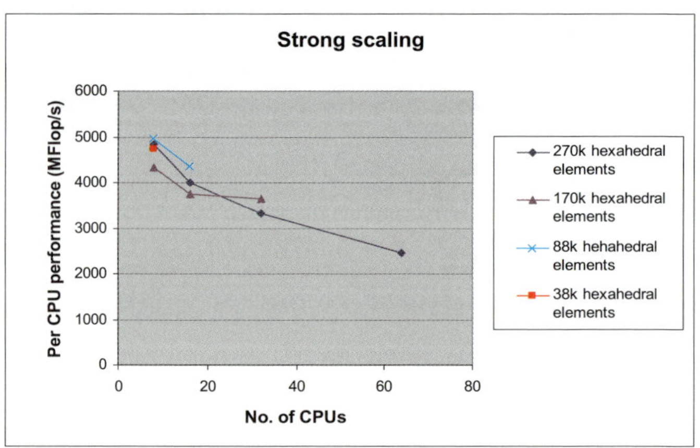

Fig. 2. Strong scaling of BLIS on NEC SX-8

3.2 Performance Comparison

Performance comparison between AZTEC and BLIS for two moderate sized problems is listed in Table 2.

AZTEC has a very rich choice of preconditioning methods and cannot be directly compared to BLIS which is still under development. Hence, no comparison of the speed of convergence or the time taken to solve the problem is included here. Jacobi preconditioning was used in both the solvers. It is

Table 2. Performance comparison of AZTEC and BLIS

No. of unknowns	No. of CPUs	Per CPU perf. (MFlop/s) AZTEC	BLIS	Factor faster
40260	8	625.3	3116.4	4.98
1143280	32	638.5	3326.9	5.21

also to be noted that though the numerical robustness of ILU preconditioning in AZTEC is far superior that Jacobi preconditioning, the floating point performance is disappointing.

The performance of AZTEC stagnates at around 650 MFlop/s on the NEC SX-8 when the native row oriented data structure is used to store the sparse matrix. The reason for such a low performance is average vector length as the innermost vectorizing loop is too short to fill up the vector pipes. BLIS performs at around 4 GFlop/s when there are above 10k finite elements partitioned per processor (around 40k unknowns per processor). This can further be improved with reducing the global synchorization needed in the algorithm and also by overlapping computation with communication where ever possible [16].

3.3 Future Work

BLIS presently works with applications that have 4 unknowns to be evaluated per grid point. This will further be extended to handle any number of unknowns. BiCGSTAB algorithm is presently available and more krylov subspace algorithms for solving symmetric and unsymmetric linear systems will be included. Overlapping communication and computation at different places in such algorithms has to be extensively looked into which is a key step towards achieving reasonable scaling for the solver. Preconditioning is the heart of any iterative solver and robust preconditioning techniques (multilevel, factorization with coloring) will be investigated next.

4 Summary

We looked into the reasons behind the dismal performance of most of the public domain sparse iterative solvers on vector machines. Then, we introduced the Block-based Linear Iterative Solver (BLIS) which is currently under development targeting performance on vector architectures. Results show promising performance on the NEC SX-8 supercomputer. Robust preconditioning is still missing and is the future direction of this project.

5 Acknowledgements

The authors would like to thank Rainer Keller of the 'High Performance Computing Center Stuttgart (HLRS)' for his helpful advice on using Valgrind and Holger Berger and Stefan Haberhauer of 'NEC - High Performance Computing Europe' for the constant technical support.

References

1. Oliker, L., Canning, A., Carter, J., Shalf, J., Ethier, S.: Scientific computations on modern parallel vector systems. In: Proceedings of the ACM/IEEE Supercomputing Conference (SC 2004), Pittsburgh, USA (2004)
2. Oliker, L., Canning, A., Carter, J., Shalf, J., Skinner, D., Ethier, S., Biswas, R., Djomehri, J., van der Wijngaart, R.: Evaluation of cache-based superscalar and cacheless vector architectures for scientific computations. In: Proceedings of the ACM/IEEE Supercomputing Conference (SC 2003), Phoenix, Arizona, USA (2003)
3. Rabenseifner, R., Tiyyagura, S.R., Müller, M.: Network bandwidth measurements and ratio analysis with the hpc challenge benchmark suite (hpcc). In Martino, B.D., Mueller, D.K., Dongarra, J., eds.: Proceedings of the 12th European PVM/MPI Users' Group Meeting (EURO PVM/MPI 2005). LNCS 3666, Sorrento, Italy, Springer (2005) 368–378
4. http://www.es.jamstec.go.jp/esc/eng/.
5. http://www.teraflop-workbench.de/.
6. Tuminaro, R.S., Heroux, M., Hutchinson, S.A., Shadid, J.N.: Aztec user's guide: Version 2.1. Technical Report SAND99-8801J, Sandia National Laboratories (1999)
7. Heroux, M.A., Willenbring, J.M.: Trilinos users guide. Technical Report SAND2003-2952, Sandia National Laboratories (2003)
8. Saad, Y.: Iterative Methods for Sparse Linear Systems, Second Edition. SIAM, Philadelphia, PA (2003)
9. Saad, Y.: SPARSKIT: A basic toolkit for sparse matrix computations. Technical Report RIACS-90-20, NASA Ames Research Center, Moffet Field, CA (1994)
10. Tiyyagura, S.R., Küster, U., Borowski, S.: Performance improvement of sparse matrix vector product on vector machines. In Alexandrov, V., van Albada, D., Sloot, P., Dongarra, J., eds.: Proceedings of the Sixth International Conference on Computational Science (ICCS 2006). LNCS 3991, Reading, UK, Springer (2006)
11. Im, E.J., Yelick, K.A., Vuduc, R.: Sparsity: An optimization framework for sparse matrix kernels. International Journal of High Performance Computing Applications **18(1)** (2004) 135–158
12. Agarwal, P., et al.: ORNL Cray X1 evaluation status report. Technical Report LBNL-55302, Lawrence Berkeley National Laboratory (May 1, 2004)
13. Nakajima, K.: Parallel iterative solvers of geofem with selective blocking preconditioning for nonlinear contact problems on the earth simulator. GeoFEM 2003-005, RIST/Tokyo (2003)
14. Jones, M.T., Plassmann, P.E.: Blocksolve95 users manual: Scalable library software for the parallel solution of sparse linear systems. Technical Report ANL-95/48, Argonne National Laboratory (1995)
15. Tuminaro, R.S., Shadid, J.N., Hutchinson, S.A.: Parallel sparse matrix vector multiply software for matrices with data locality. Concurrency: Practice and Experience **10(3)** (1998) 229–247
16. Demmel, J., Heath, M., van der Vorst, H.: Parallel numerical linear algebra. Acta Numerica **2** (1993) 53–62

Visualization: Insight on Your Work
Real-time and Large-scale Visualization

Pascal Kleijer

NEC Corporation, HPC Marketing Promotion Division,
1-10, Nisshin-cho, Fuchu-shi, Tokyo, 183-8501, Japan
`k-pasukaru@ap.jp.nec.com`

Summary. Visualization is an important part of numerous simulation workflows. It helps users intuitively discover artifacts in their data, because it directly makes use of one of human's fundamental sense: "vision". Also, it does not require as much effort as raw data analysis. There are several usable workflows depending on current needs, each one using a different visualization approach. With the increasing power of computation machines, now easily reaching the teraflops barrier, the amount of data to handle might be a bottleneck for commodity rendering hardware. Visualization therefore needs specific tools in order to cope with this issue. This paper provides an overview on the different workflows with their strengths and weaknesses as well as their best usage. It will also focus on concrete examples of large-scale visualization and finally present one application suited for a particular workflow.

Key words: visualization, remote rendering, workflow, real-time, large-scale, simulation, steering, tracking.

1 Introduction

Nowadays, commodity hardware for graphical rendering (such as proposed by NVIDIA [1] or ATI [2]) is powerful and feature-rich. This has led to the development of powerful tools to visualize in real-time the various data resulting from simulations. It has democratized the usage of rendering tools for scientific visualization or for publication visualization. Scientific visualization tries to represent the simulation results as close as possible to reality so that analysis can be performed accurately, where publication visualization is more interested in producing an aesthetically pleasing image for non-experts, see on the right of Fig. 1 as example [3]. In the latter, accuracy isn't important since the image just helps to illustrate the idea.

Whatever visualization style is used, the data source generally remains the same. Even with such democratization, limitations soon appear as the amount

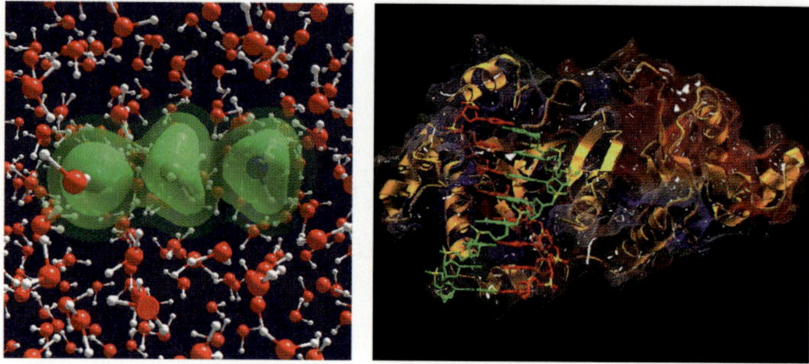

Fig. 1. Scientific vs. Publication Visualization

of data increases. High-end models of commodity hardware can usually support up to 1 gigabyte of dedicated memory; this isn't enough for datasets that are several gigabytes in size. Such large datasets are no longer unusual, when teraflops computational systems with several gigabytes to terabytes of memory are used.

2 Large-scale Visualization Needs

An efficient approach is required for visualizing an enormous amount of distributed data, which is produced by parallel simulations that are conducted on large computational systems or in grid computing environments. Conventionally, visualization refers to post-processing simulation results on a user's terminal after the numerical simulation has been completed. The simulation results have to be output to the computational server's disk and transferred to the user's terminal before they are processed. Problems appear when visualizing a large volume of data with conventional post-processing. For one thing, a large volume of distributed numerical data has to be combined into one batch of organized data and then transferred over the network. Moreover, a large amount of disk space is necessary to store the data somewhere in the computing environment and on the user's terminal. Finally, a large amount of memory space is necessary to manipulate the data on the user's terminal.

3 Workflow

Regardless how visualization is performed, the general workflow can be cut down to four major phases as shown in Fig. 2.

In the first phase, the simulation performs various calculations. This is either a single pass or an iterative process to reach the next step; it is highly dependent on the problem to be solved. It might also be a simple post-processing

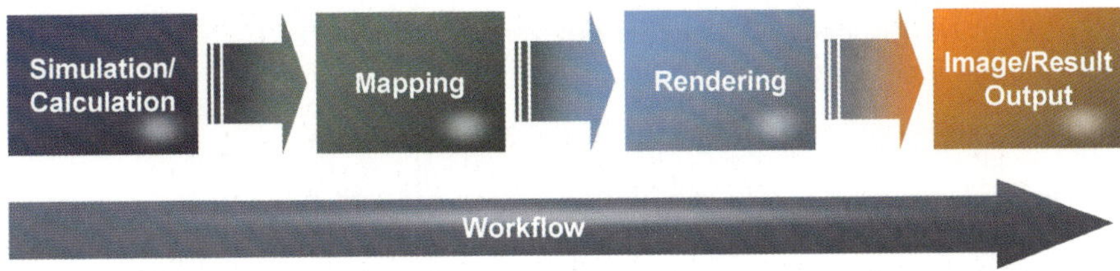

Fig. 2. Workflow Phases

application which loads and aggregates data. The result is then carried over to a mapping engine that has the task to create graphical objects. All graphical objects are then provided to the rendering engine which produces an image. The image is either saved somewhere on a hard disk resource or transferred to a user's terminal for direct visualization. Between each successive phase there is data movement of some variety.

In the current discussion, we assume the usage of a two-peer system. On one end the simulation is running on a computational server and on the other end the user's machine or a hard disk resource is used. Depending on the mapping and rendering phase distribution, the resulting approaches have their strengths and weaknesses.

3.1 Approach A

This approach is the classical workflow used for the large majority of the cases. The simulation runs on its dedicated machine and the rest of the workflow it is handled by the second machine, see Fig. 3.

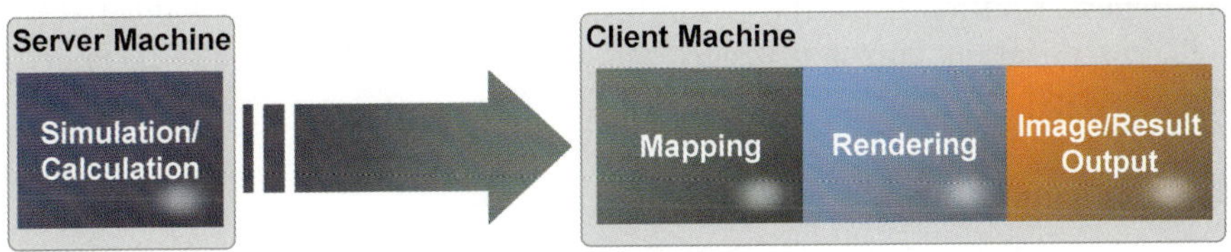

Fig. 3. Approach A

Both the mapping and rendering phases are assigned to the client machine. The data to be transferred are the computation results. The advantage of this approach is that the server can be dedicated to the numerical simulation. However, it is almost impossible for the network to transfer the entire volume of raw data every time-step, and for the user terminal to visualize it. Some kind of mechanism is necessary to extract the part of the data that is actually used for the visualization. Moreover, important information contained in the raw data might be dropped and visualization results, such as the tracer calculation, can become inaccurate.

Pros

- Dedicated machine. The simulation server can run fully on its dedicated resource. The client machine is generally a powerful rendering server like an SGI or high end graphical workstation with dedicated 3D hardware.
- High level of interactivity. Since the rendering is done on the same machine as the visualization, high interactivity is expected.
- Classical workflow: plenty of tools. The number of tools available either freely or commercially comes in truck loads. Each specific domain has its list of applications. Some well known are AVS [4], RasMol [5], EnSight [6], FieldView [33], etc.

Cons

- Must wait until the simulation is finished. There is no way to know if the simulation has run correctly, the sequence of operations are such that only post-processing visualization is possible.
- Network bandwidth over-utilization. The size of the problem is generally big enough to put a large burden on both the storage and the network. If each time step produces an extra gigabyte of data, it will take time to transfer. Another solution is to use global file systems, but it restricts the workflow to a local usage.
- Down scale of data before visualization. In most cases it is necessary to down scale the data either for the network or for the rendering engine. This creates inaccuracies in the visualization.

This approach is well suited in the case the data must be stored anyway for future usage, astrology or oceanic studies are typical examples. It is also the defacto solution when the computation resource can only be dedicated to, well, computation; this is typically the case for large systems like the Earth Simulator [7].

3.2 Approach B

This approach is a hybrid workflow approach used in some rare cases. The mapping process is assigned to the server side whereas the rendering process is assigned to the client side, see Fig. 4. The data transfer is based on 3D graphical objects.

The advantage of this approach is that the mapping phase can utilize the computation performance of the server and the rendering process is accelerated by special graphics hardware integrated in the user terminal. However, because the number of 3D graphical objects is usually proportional to the simulation scale, there is the same network load problem as in Approach A. Furthermore, some visualization techniques (such as volume rendering) do not use graphical objects and have to be performed on the server side after all.

Fig. 4. Approach B

Pros

- Dedicated machine. The simulation server can run fully on its dedicated resource. The client machine is generally a powerful rendering server like an SGI or high end graphical workstation with dedicated 3D hardware.
- Good level of interactivity. When it comes to just changing the viewpoint, lighting, etc. this approach is very efficient and gives an excellent interactivity level. However, when switching between visualization tools (tracer, isosurfaces, etc.), interactivity drops sharply.
- Easy support of multi-users. Once the 3D scene is generated, multiple users can share it on different machines.

Cons

- Down scale of data before visualization. In most cases it is necessary to down scale the 3D graphical objects, the most obvious is polygon reduction in tools such as isosurfaces. This might create inaccuracies in the visualization.
- Puts a heavy load on the network. The 3D graphical objects can quickly reach large sizes if proper downscaling is not carried out.
- Not all rendering techniques are possible. Some tools such as volume rendering are just impossible to use with this configuration.

Such system can be used with simulations that can easily define Meta 3D objects instead of directly defining polygons and vertices. For example, molecular science with large atomic models can easily use this approach. It also allows easy collaboration with a common scene shared among members, useful if the client machine has 3D hardware available. There are systems implementing this approach such as RAVE [8].

3.3 Approach C

This is an alternative to the client side rendering as proposed in approach A. It is used in some cases. This solution, we will see, offers the best response to tera- or peta-scale problems. Here, both the mapping and rendering phases are assigned to the server side, see Fig. 5. The data to be transferred are 2D images. The volume of transferred data depends only on the image size

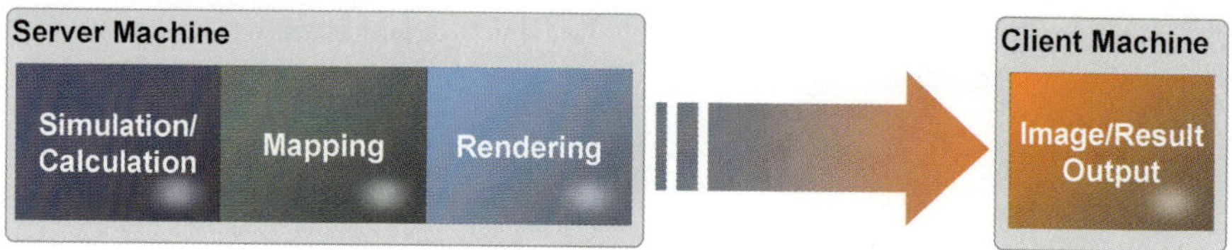

Fig. 5. Approach C

and the number of colors used. It is independent from the simulation scale. Furthermore, using image compression techniques like JPEG [9], PNG [10] or even streaming, traffic volume can be dramatically reduced.

In real-time systems, the transfer of images between the server and the client can be streams based rather then unique images. This can reduce latency, thus increase overall network speed, however encoding/decoding overhead will appear.

Pros

- Maximum resolution. The data can be directly shared with the mapping and rendering engine. It is possible to do visualization without losing a single bit of information.
- Solely uses existing hardware. It does not require other hardware then the computation server. The client can be installed on any system, even thin devices such as Personal Digital Assistant or mobile phones.
- Only the image size influences network load. The client resolution dictates the size of the images which in turn will define the necessary bandwidth. With current compression techniques for on demand video or static images, the bandwidth requirements can be very low.
- Easy concurrent and post-processing visualization. The model enables concurrent visualization since the visualization is directly coupled to the simulation.

Cons

- Lower interactivity. The latency between the user actions and the image refresh can take some time depending on the complexity, data size and the simulation step length.
- Uses computational power for rendering. It might be an issue when computational resources are used to do other tasks then calculations.
- No 3D hardware support. Rendering might not or in very few cases make us of 3D hardware acceleration. A computational server does in general not support graphical hardware.

Server-side visualization is one solution for large-scale visualization. It provides a way to follow the tera- or peta-scale computation power without creating visualization bottlenecks. Especially, by conducting concurrent visualization on the server, the storage and transfer of data resulting from the simulation can be minimized, thus making possible efficient visualization that is tailored to high-end computing.

4 Examples

The following examples are all existing research and industry problems using server side visualization. All visualization tools used in the examples are developed by us.

4.1 Ab Initio Calculations

Based on first-principles calculations, Keihanna Interaction Plaza Inc. [11] and Doshisha University [12] are involved into the study of the electronic structure and structural properties at the interface between an organic molecule and a metal surface. Interface phenomena occurring with organic-metal systems are often complex and require a robust simulation methodology. At present, they use a recently developed parallelization scheme for plane-wave pseudo potential method, based on the density functional theory within the local density approximation to calculate the electronic ground state of large systems efficiently [13]. The present scheme involve the solution of the Kohn-Sham equation [14] iteratively using residual minimization techniques, which are more suited for parallelization of large atomic systems than other minimization schemes. After atomic relaxation, the stable interface structure converges to zero force. Once stability is reached, detailed electronic properties are analyzed.

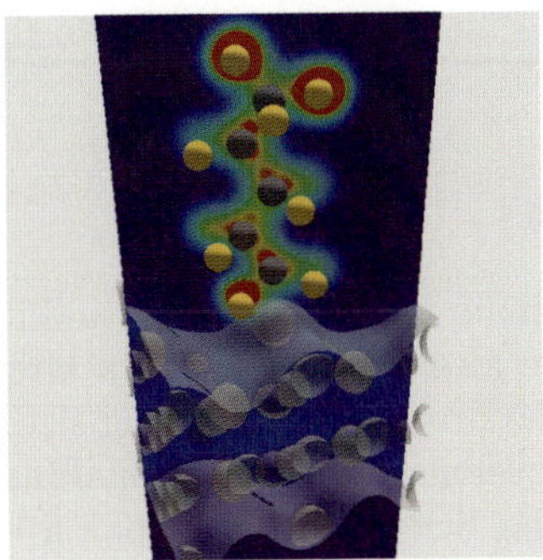

Fig. 6. Relaxed charge density distribution at the organic-metal interface

Figure 6 represents the stable interface charge density distribution obtained after atomic free relaxation of a metal-organic interface using a fluorine based molecule absorbed on a metal surface. The rendering uses a contour plane, with particle atoms for the molecule and semi-transparent isosurfaces for the metal surface.

4.2 Medical Biofluids

CCRLE [15] and University Sheffield [16] are involved in the research of new techniques for biofluids simulations [17]. Cardiovascular diseases annually claim the lives of approximately 17 million people worldwide. Atherosclerosis is one particular disease which causes the formation of deposits (plaque) on the inner lining of a vessel. It can lead for example to stroke, heart attack, eye problems and kidney problems. A secondary concern associated to the disease is the flow disturbances caused by the narrowing of the vessel lumen (vessel stenosis). Accurate scientific visualization is primordial in this project: it must help medical doctors support their decision of treatment or surgery. The work involves simulation of blood clotting in vessels due to a stenosis.

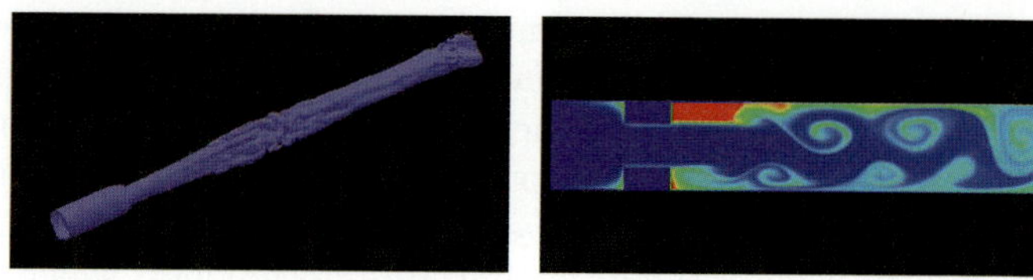

Fig. 7. Idealized Artery Stenosis

Figure 7 represents an opaque isosurface showing turbulences occurring after a 75% stenosis. The second picture, with the same turbulent flow but with a contour plot, shows blood starting to clot in the red regions just behind the stenosis.

4.3 Molecular Science

Molecular science is a domain in full expansion and its requirements are huge. Visualization is of course also an area where a large number of tools exist. However, to cope with the problems' increasing size, new tools are and must be developed. We have developed for the National Institute of Informatics [18] and Institute for Molecular Science [19] in Japan a visualization tool capable of large-scale massive distributed data handling over Grid enabled systems in the framework of the NAREGI Project [20].

Figure 8 presents from left to right, a ten million liquid water molecule cube, a virus core of two million atoms and a close up zoom on the bonding of molecules in a water solvent with semi-transparent isosurfaces.

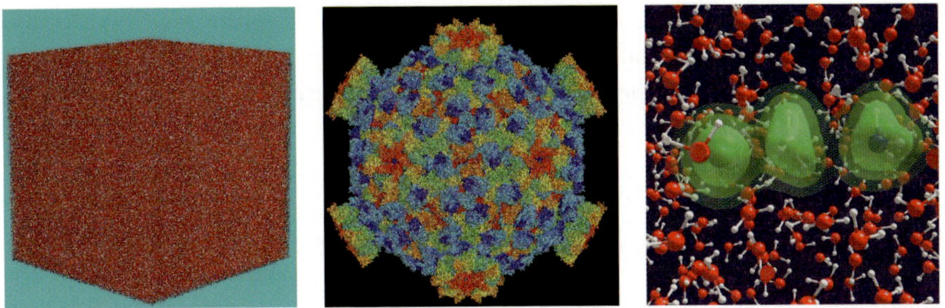

Fig. 8. Different Molecular Scene Views

4.4 Space Weather Forecasting

A rather less known topic, but very important, is space weather forecasting. This domain is crucial for satellite owners. If correctly predicted and announced in advance, the owners can take preemptive measures to protect their fragile multi-million dollar assets in space. For example, in case of solar flares, early prediction of the impact on earth neighboring space can have a major effect on business (telecom disruption or satellite damage).

In collaboration with National Institute of Information and Communications Technology in Japan [21] we helped to develop their workflow. Real-time solar wind data is collected by the ACE satellite (Advanced Composition Explorer) [22]. It is then computed with a MHD (Magneto-Hydro-Dynamic) simulation program [23]. Finally, images and videos are outputted representing different quantities such as magnetic fields behavior or particular pressure distribution.

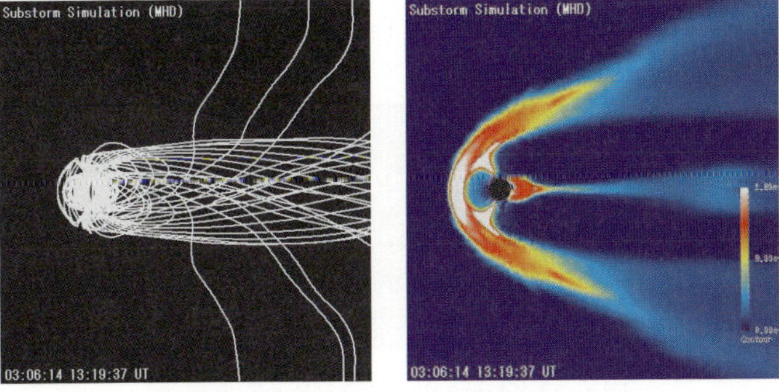

Fig. 9. Magnetic Field and Particle Pressure around Earth

Figure 9 represents a contour line plot of the magnetic field on the left side and a colored contour plane of the particle pressure on the right side. The small black dot in the center left on the image is the Earth.

4.5 Meteorology

One of the most common large-scale simulations is weather forecasting. This traditional area is very demanding in regards to visualization because the information must be distributed to a wide group of users, who might not necessarily be experts. In collaboration with the Bureau of Meteorology in Australia [24], a project was set up to focus on real-time visualization of weather forecasting and atmospheric simulations over the Australian continent, see Fig. 10.

Fig. 10. The Australian Continent

Numerous simulations and visualizations have been conducted, including but not limited to, visualization of relative humidity, pressure isobars, temperature isobars or wind velocity, see Fig. 11.

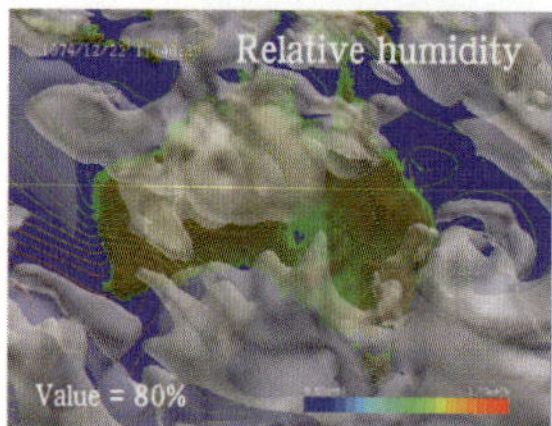

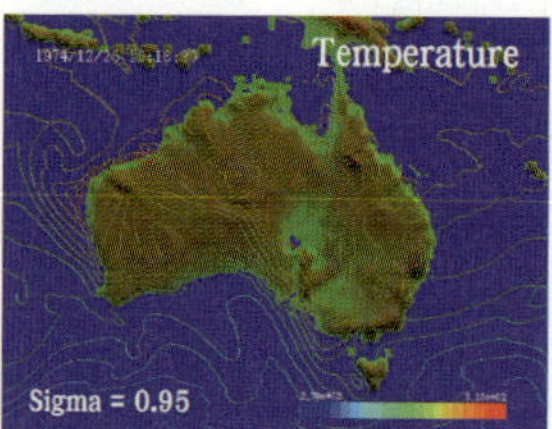

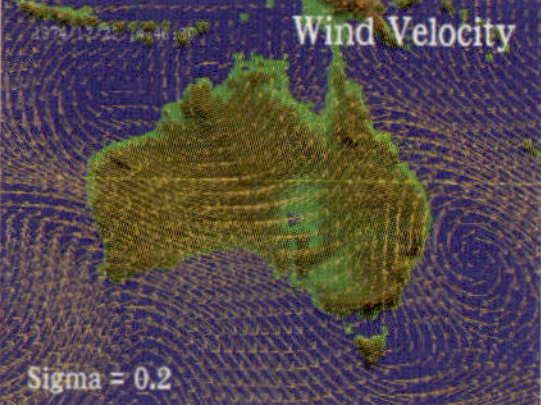

Fig. 11. Australian Weather Forecasts

4.6 Vehicle Design

Nissan Motors [25], like many in the automobile industry, has always been a great consumer of large-scale computational resources. Many aspects of a car design require large-scale simulation such as fluid flow analysis for aerodynamics. Their workflow does of course need visualization.

We optimized their workflow, Fig. 12, to achieve a full cycle in one day instead of seven days with their previous approach. To produce this, the visualization is directly embedded into the CFD simulation. Numerous videos and images are produced to analyze various aspects of car design. The outputs are made directly available when the simulations have completed.

One of Nissan's flagship cars, the "Fairlady Z" model, was designed using this workflow, Fig. 13. Numerous models of high levels of detail were used to produce movies and pictures of the car in the design phase.

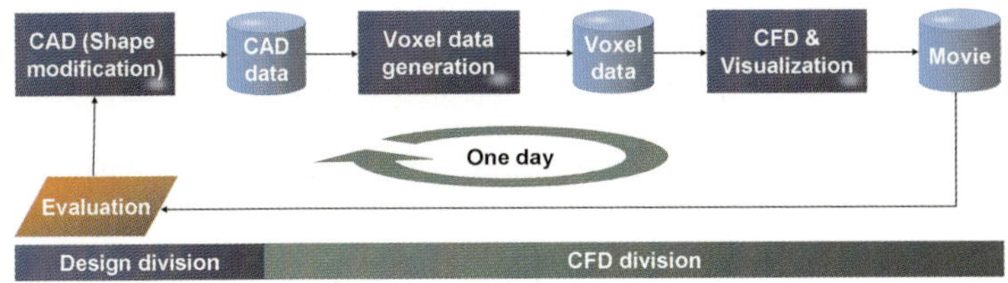

Fig. 12. Nissan Workflow

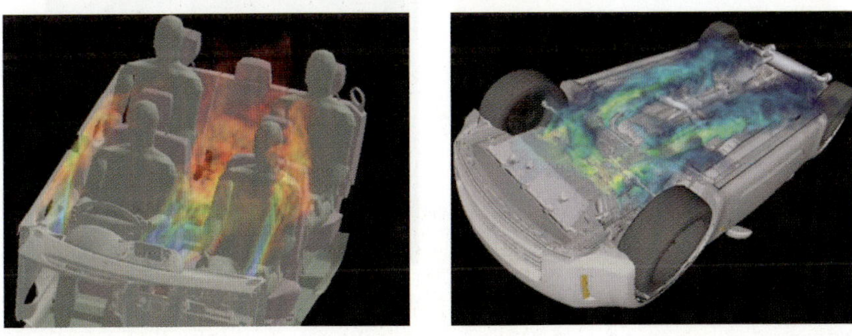

Fig. 13. Temperature distribution in the cabin and air flow under the vehicle with volume rendering

4.7 Sport

The Institute of Physical and Chemical Research RIKEN in Japan [26] studied the pitch of Japanese professional baseball pitcher Matsuzaka. He has a particular way of throwing his pitches. This study [27] was undertaken to analyze how his Gyro-ball breaking pitches work out, Fig. 14. This is a typical fluid dynamics computation problem around a round shaped object.

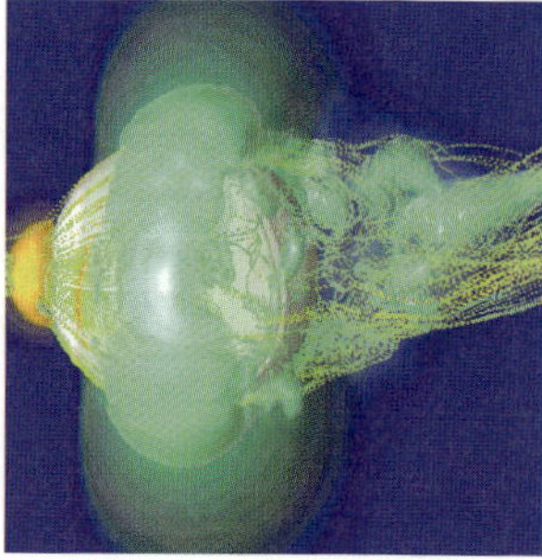

Fig. 14. Mr. Matsuzaka and his breaking pitch

Still in the fluid flow visualization, another piece of work with the Nakahashi Laboratories of Tohoku University in Japan [28] tried to reproduce the behavior of a golf ball, Fig. 15. Golf is an important sport in Japan's business world; to assist players in winning, the inner secrets of golf need to be revealed.

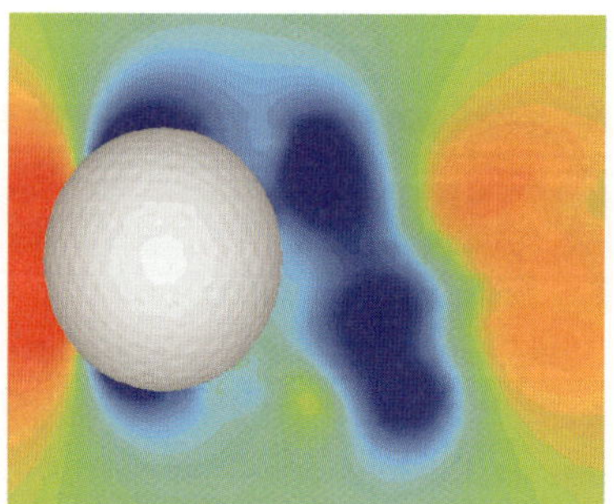
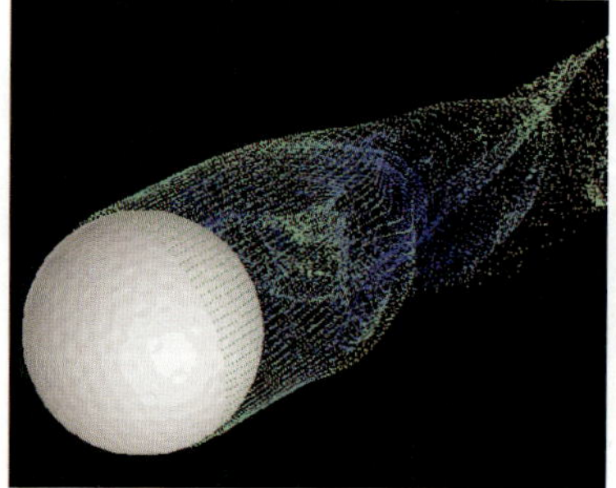

Fig. 15. Golf ball with contour plane and tracers

5 RVSLIB

NEC as a High Performance Computing platform provider, such as the SX series, is interested in supporting visualization workflows. In order to provide a wide range of solutions to its customers, NEC has developed over the last

15 years a special visualization application called **RVSLIB** [29] [30], which stands for **R**eal-time **V**isual **S**imulation **Lib**rary. Unlike most other tools, this particular one is tailored to run in either post-processing or concurrently with simulations. This application applies the approach C principle (server side rendering), thus seeking to help the problem with massive data for large-scale computing environments. It can sustain high performance computation on both scalar and vector architectures despite the fact it uses software rendering.

5.1 Configuration

The environment can be setup in two different ways as shown in Fig. 16. Depending on the needs, security level, global network or workflow, one or both solutions can be installed.

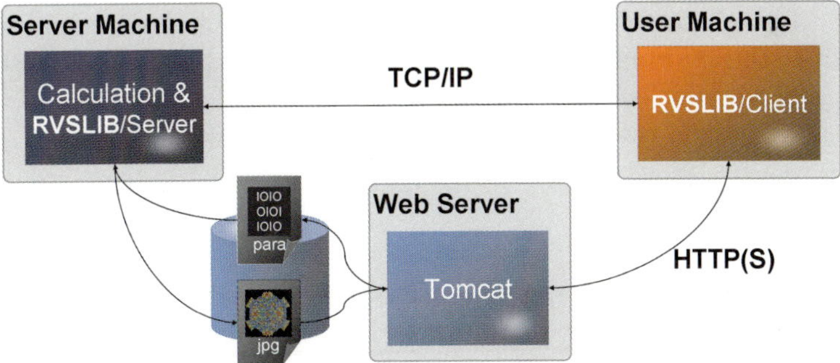

Fig. 16. General Configuration

RVSLIB uses server-client architecture, consisting of **RVSLIB**/Server and **RVSLIB**/Client. Both components can either be installed on the same machine or on different machines connected by internet or intranet. Communication between the modules is either by direct connection or using a web service relay for Tomcat [31]. In the first case, the connection is done by using the Telnet protocol and direct TCP/IP (Transmission Control Protocol/Internet Protocol) socket streaming. For the latter, the client connects by standard HTTP(S) (Hyper Text Transfer Protocol (over SSL)) requests to the service. The service will relay the orders via files to the **RVSLIB**/Server. The web service based solution offers a better secure environment and allows users outside the firewalls to use the system safely.

5.2 Principle

The principle is simple. The rendering is done on the computational server as defined in the server side rendering. The information transits over the intranet or internet to the user's terminal as in Fig. 17.

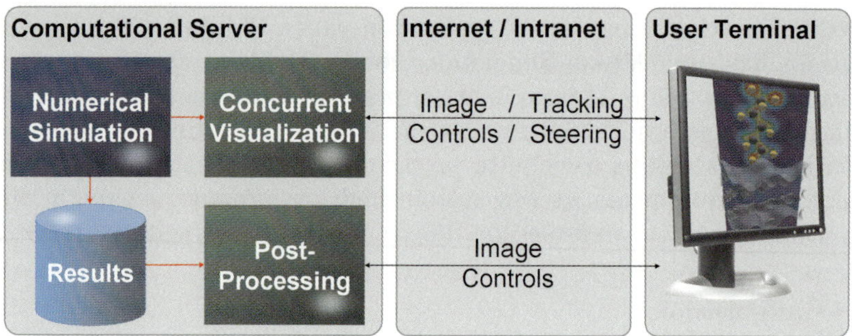

Fig. 17. RVSLIB Principle

The web server shown in Fig. 16 is located on the network part and is not shown in Fig. 17. Depending on the type of visualization, concurrent or post-processing, the information communicated differs slightly. In the first case, tracking information and steering data can be added to the normal image-controls flow, which is the only flow available in the latter.

5.3 Integration

The integration is rather trivial. Only the server code needs to be modified as shown in Fig. 18.

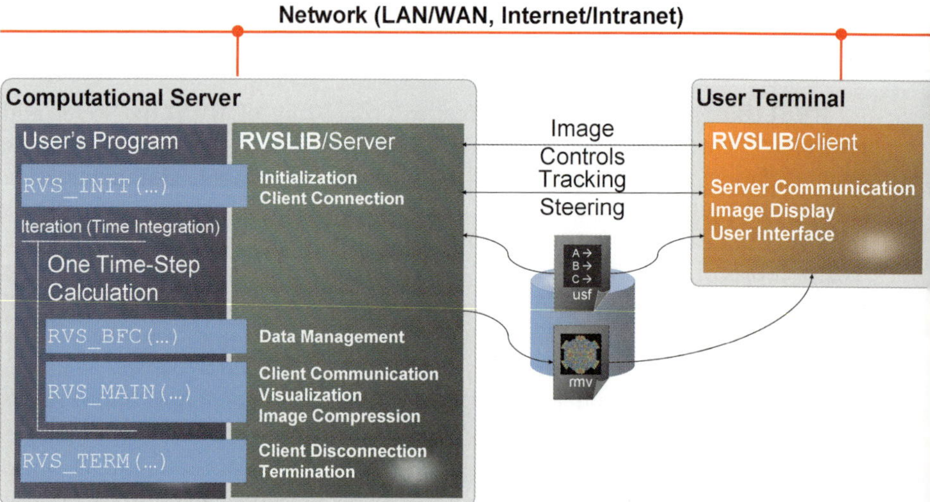

Fig. 18. Server Integration

RVSLIB/Server supports a set of FORTRAN/C subroutines to be incorporated into the user application's code. A certain number of these routines are mandatory; however most will be optional and depending on what is necessary for the specified code. **RVSLIB**/Client enables the user to display an image created by an **RVSLIB**/Server and to steer simulation parameters through a user interface. Automatic control of the **RVSLIB**/Server or/and the **RVSLIB**/Client is possible with scenario files. Videos, in proprietary format, can be generated on the **RVSLIB**/Server on the fly, and then visualized on the **RVSLIB**/Client.

5.4 Main Features

Data Type

The lattice data format and analytical results format can be delivered to the library by a set of routines. These routines will request the parameters to be formatted in a specific way, for example:

```
Lattice:
    X(MAXI, MAXJ, MAXK),
    Y(MAXI, MAXJ, MAXK) and
    Z(MAXI, MAXJ, MAXK)

Analytical Values:
    SCAL(MAXI, MAXJ, MAXK, NS) and
    VECT(MAXI, MAXJ, MAXK, 3, NV)
```

Where NS and NV are the number of items for each grid cell. MAXI, MAXJ, and MAXK are the maximum numbers of lattices in each direction when the lattice space defined by (I, J, K).

The library supports 4 different formats: Single / Multi-Block BFC Grid, Unstructured Tetrahedron Grid, Unstructured Hexahedron Grid and Particle Data. Grid data are mutually exclusive but can all be used in combination with particle data.

Tracking/Steering

Computational results are visualized simultaneously with the on-going simulation. The simulation parameters can be changed interactively without disrupting the simulation.

Plug-in/Plug-out

The communication between **RVSLIB**/Server and **RVSLIB**/Client can be suspended or re-established at any time. Even for a time-consuming simulation, it is necessary to establish the network connection between the server and the client only when monitoring of the on-going simulation is necessary, resulting in the efficient use of the computing resource. Even if a sudden communication problem occurs, the numerical simulation will continue, the **RVSLIB**/Client has only to reconnect to the server module.

Scenarios

The scenario automates the simulation and produce stunning output without human interaction. Most visualization parameters can be controlled by the scenario; this enables it to generate a full movie with camera motion, control changes, etc without interaction. Launch it in the evening and in the morning visualize the fresh outputted video ready for distribution.

Videos and Images

You can generate a movie manually or with the help of a scenario. The movie format is proprietary to **RVSLIB** but ensures maximum image quality; it can also be converted to MPEG [32] movies with a Windows conversion tool provided with the distribution. It is also possible to save a succession of JPEG [9] images and build (with a 3rd party tool) a video in any format.

Versatile Visualization Functions

Numerous visualization functions for any kind of rendering are available: Object, Contour Plane, Contour Line, Lattice Grid, Vector, Tracer, Stream Line, Isosurfaces, Volume Rendering, Particle, Color bar, Title...

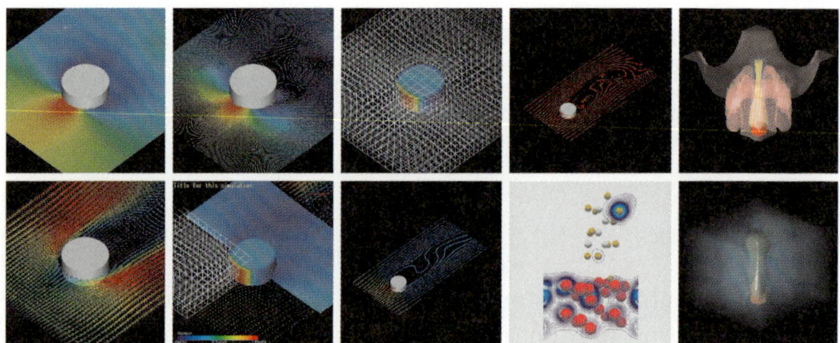

Fig. 19. Visualization Function Examples

6 Conclusion

In this report we have explained the different visualization workflow approaches existing for two-peer systems. It showed some of the strengths and weaknesses of each approach as well as the best application. It followed with different illustrated examples of large scale massive data visualization all using a similar approaches, server side visualization all using a similar approach: server side visualization, which was developed by NEC. Finally we introduced one of our flagship solutions: **RVSLIB**.

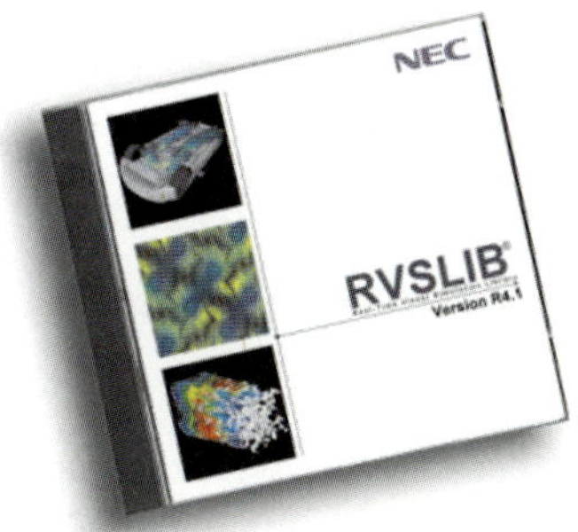

Fig. 20. RVSLIB CD case

For future inquiries about the **RVSLIB** product or visualization related solutions please contact:

HPC Marketing Promotion Division
NEC Corporation
Fax: +81 - (0)42/333.6382
E-mail: **support@rvslib.jp.nec.com**
Home Page: `http://www.sw.nec.co.jp/APSOFT/SX/rvslib_e`

References

[1] NVIDIA, Graphical Processor Technology manufacturer.
`http://www.nvidia.com`
[2] ATI, Graphical Processor Technology manufacturer.
`http://www.ati.com`

[3] DNA binding to TnsA, Image courtesy F. Dyda (NIDDK) from NIH Biowulf.
`http://biowulf.nih.gov/apps/povray/povray_examples.html`

[4] AVS (Advanced Visual Systems), visualization softwares.
`http://www.avs.com`

[5] RasMol, Freeware Molecular Visualization.
`http://www.umass.edu/microbio/rasmol`

[6] EnSight, FEA and CFD post-processing visualization and meshing tool.
`http://www.ensight.com`

[7] Japan Agency for Marine-Earth Science and Technology Earth Simulator Center.
`http://www.es.jamstec.go.jp/esc/eng`

[8] I.J.Grimstead, N.J.Avis and D.W.Walker, "RAVE: The Resource-Aware Visualization Environment". Concurrency and Computation: Practice and Experience. 2006; vol: 1-15.
`http://www.wesc.ac.uk/projects/rave`

[9] Joint Photographic Experts Group (JPEG):
`http://www.jpeg.org`

[10] Portable Network Graphics (PNG) - W3C Recommendation - Second Edition, November 2003.
`http://www.w3.org/TR/PNG`

[11] Kyoto Prefecture Collaboration of Regional Entities for the advancement of Technological Excellence, Japan Science and Technology Agency, 1-7 Hikaridai, Kyoto 619-0237, Japan.

[12] Department of Chemical Engineering and Materials Science, Doshisha University, 1-3 Kyotanabe, Kyoto, 610-0321, Japan.

[13] T. Tamura, G. Lu. R. Yamamoto, M. Kohyama, S.Tanaka and Y. Tateizumi, Model. Simu. Mater. Sci. Eng., 12 945 (2004).

[14] W. Kohn and L. J. Sham Phys. Rev. 140, A1133 (1965).

[15] C&C Research Laboratories, NEC Europe Ltd., St.Augustin, Germany.
`http://www.ccrl-nece.de`

[16] Medical Physics and Clinical Engineering, University of Sheffield, Royal Hallamshire Hospital, Glossop Road, Sheffield, S10 2JF (UK):
`http://www.shef.ac.uk/dcss/medical/medicalphysics/medical-physics`

[17] JÃűrg Bernsdorf, Sarah Harrison, D. Rodney Hose, P.V. Lawford and S. M. Smith, "Concurrent Numerical Simulation of Flow and Blood Clotting Using the Lattice Boltzmann Technique", proceedings of the IEEE Computer Society Press, ICPADS 2005.

[18] National Institute of Informatics, 2-1-2 Hitotsubashi, Chiyoda-ku, Tokyo, Japan.
`http://www.nii.ac.jp/index.shtml.en`

[19] Institute for Molecular Science, Myodaiji Area: 38 Nishigo-Naka, Myodaiji, Okazaki 444-8585, Japan.
`http://www.ims.ac.jp`

[20] National Research Grid Initiative (NAREGI).
`http://www.naregi.org/index_e.html`

[21] National Institute of Information and Communications Technology, NICT.
`http://nict.jp`

[22] ACE (Advanced Composition Explorer) satellite.
`http://www.srl.caltech.edu/ACE`

[23] Den, M., et al. (2006), Real-time Earth magnetosphere simulator with three-dimensional magnetohydrodynamic code, Space Weather, 4, S06004.

[24] Australian Gouvernement, Bureau of Meteorology.
http://www.bom.gov.au

[25] Nissan Motors.
http://www.nissan.co.jp

[26] Institute of Physical and Chemical Research RIKEN.
http://www.riken.go.jp/engn

[27] Computer Visualization Contest 2000.
http://accc.riken.jp/HPC/CVC2000

[28] Nakahashi Laboratories of Tohoku University.
http://www.ad.mech.tohoku.ac.jp

[29] T. Takei, J.Bernsdorf, N. Masuda, and T. Takahara, "Lattice Boltzmann Simulation and Its Concurrent Visualization on the SX-6 Supercomputer", In 7th International Conference on High Performance Computing and Grid in Asia Pacific Region, 2004.

[30] Real-time Visual Simulation Library (RVSLIB).
http://www.sw.nec.co.jp/APSOFT/SX/rvslib_e

[31] Apache Jakarta Project - Tomcat.
http://jakarta.apache.org/tomcat

[32] Moving Picture Experts Group (MPEG).
http://www.chiariglione.org/mpeg

[33] FieldView, Post-processor and visualization software for computational fluid dynamics.
http://www.ilight.com

Computational Fluid Dynamics

Control of Turbulent Boundary-Layer Flow Using Slot Actuators

Ralf Messing, Ulrich Rist[1], and Fredrik Svensson[2]

[1] Institut für Aerodynamik und Gasdynamik (IAG), Pfaffenwaldring 21, 70550 Stuttgart, Germany [last name]@iag.uni-stuttgart.de
[2] NEC High Performance Computing Europe GmbH, Heßbrühlstraße 21B, 70565 Stuttgart, Germany fsvensson@hpce.nec.com

This paper describes the advances made by the collaboration between the Institut für Aerodynamik und Gasdynamik and the Teraflop Workbench. The target was to enable new research on the large SX-8 system at HLRS.

1 Introduction

For the case of flows over solid surfaces, the separation of the boundary layer causes large energy losses which in turn strongly affect the aerodynamic loads in terms of lift loss and drag increase. Therefore, there is a strong need to delay or even eliminate the occurrence of flow separation. Regarding commercial aircrafts the delay or elimination of separation of the typically turbulent boundary layer on the wing would permit higher angles of attack during landing and take-off. Using appropriate means for separation control one could even think of a high-lift system without slat (slatless wing) leading to devices with less maintenance effort and noise production.

In order to manipulate and control separated turbulent boundary layers jet actuators have been proposed injecting fluid into the boundary-layer flow by continuous or pulsed blowing. The influence of geometry and orientation of the orifices as well as the direction of the fluid jets (parallel, inclined or normal to the wall) is actually under examination in research studies. Basically, the application of these jet actuators aims at enhanced mixing rates in the boundary layer increasing momentum in the vicinity of the wall.

This paper presents a comparison of experiments conducted at the Institut für Strömungsmechanik at the Technical University of Braunschweig and direct numerical simulations done at the Institut für Aerodynamik und Gasdynamik.

1.1 Numerical Method

Details of the numerical method have been reported in various publications [1,2,4]. Therefore, the description of the numerical method can be restricted to the modifications that had to be carried out to simulate a slot actuator with steady blowing. Blowing is modeled by prescribing the steady wall-normal velocity at the wall:

$$v'(x, 0, z) = v_c \cos^3 \left(\frac{\pi r}{d} \right).$$

(1)

For spanwise slots with a slot width d_{SL} (extension in chordwise direction) and a slot length L_{SL} (extension in spanwise direction) follows :
For $z_{SL} \leq z \leq z_{SL} + L_{SL}$

$$d = d_{SL} \,, \; r = \sqrt{(x - x_{SL})^2} \,, \; r \leq \frac{d_{SL}}{2}$$

(2)

For $z < z_{SL}$

$$d = d_{SL} \,, \; r = \sqrt{(x - x_{SL})^2 + (z - z_{SL})^2} \,, \; r \leq \frac{d_{SL}}{2}$$

(3)

For $z > z_{SL} + L_{SL}$

$$d = d_{SL} \,, \; r = \sqrt{(x - x_{SL})^2 + (z - (z_{SL} + L_{SL}))^2} \,, \; r \leq \frac{d_{SL}}{2}$$

(4)

The slots have circular roundings at their lateral ends. The center of the rounding is located at $(x_{SL}, 0, z_{SL})$. In contrast to older versions of the numerical code the slots can now be rotated in the wall plane with respect to the main flow direction (skew angle β, see Fig. 1). Blowing is still perpendicular to the wall (pitch angle α, see Fig. 1).

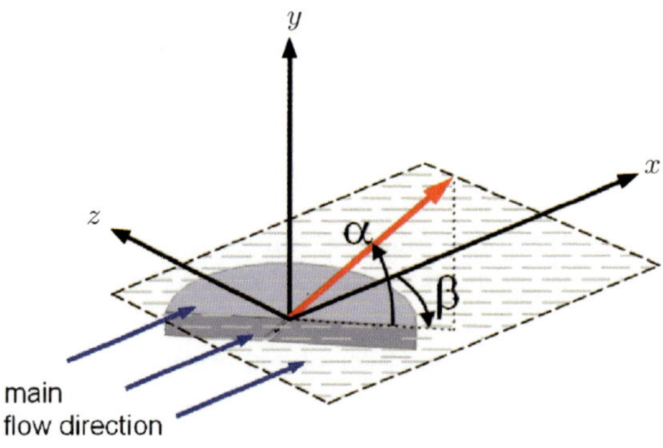

Fig. 1. Sketch of actuator for definition of pitch angle and skew angle

2 Results

2.1 Undisturbed Flow

Experimental data has been provided for a flate plate with zero-pressure gradient at a freestream velocity $U_\infty = 15m/s$ to gain some detailed insight into the effectiveness of skewed slot actuators. However, before comparing experimental and numerical results it has to be ensured that the turbulent bounday-layer flows without actuators match. To establish turbulent boundary-layer flow the numerical simulation is carried out according to the procedure in the experimental set-up. There, turbulence is triggered by an adhesive tape mounted downstream of the leading edge of the plate, and far enough upstream of the measurement station. A very similar approach is applied in the numerical set-up. By harmonic suction and blowing in a disturbance strip unsteady disturbances are excited which lead to rapid breakdown of the initially laminar flow and rapidly provide a fully-developed turbulent boundary layer downstream of the disturbance strip. To illustrate this, the skin friction coefficient is plotted over the whole integration domain in Fig. 2. The disturbance strip is located at $x = 14.34$. Downstream the wall friction coefficient c_f strongly increases due to laminar-turbulent transition and rapidly approaches the values for fully-turbulent flow. Despite the penalty of additional computational time and memory requirements this approach has been prefered because it does not suffer from somehow unphysical initial boundary conditions for turbulent flow.

Finally, mean velocity profiles and rms-profiles of the streamwise velocity component at $Re_{\delta 1} = 1855$ are compared in Fig. 3 to mutually validate experiment and numerical simulation. Agreement is quite satisfactory, even though some minor deviations are discernible in the rms-profiles near the wall.

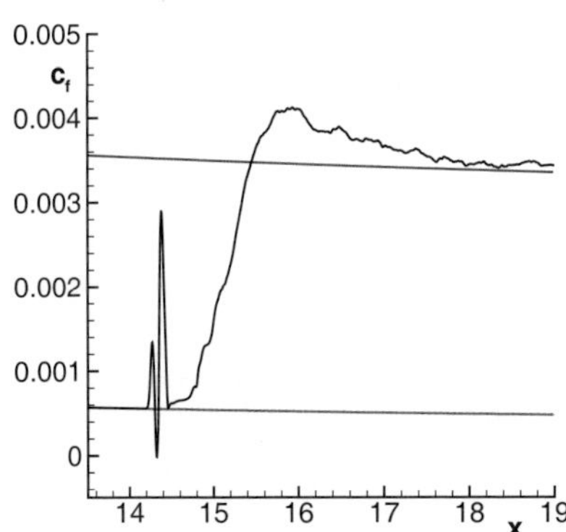

Fig. 2. Skin friction coefficient versus streamwise direction. Black line: simulation; Red line: laminar flow; Blue line: turbulent flow

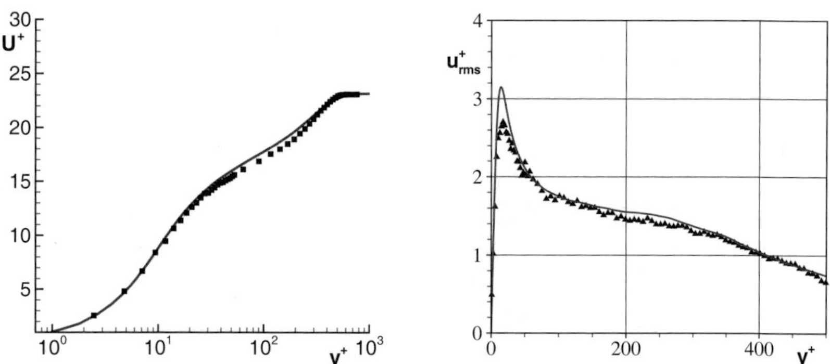

Fig. 3. Comparison of experiment (symbols) and simulation (lines) at $Re_{\delta1} = 1855$; left: streamwise mean velocity profiles; right: streamwise rms-profiles

2.2 Disturbed Flow

At the University of Braunschweig a measuring campaign was conducted to identify the most effective configuration to increase momentum near the wall. As mentioned a flate plate with zero-pressure gradient has been chosen to minimize expenses as extensive parametric studies were necessary. A slot with preferably steady blowing with a pitch angle $\alpha = 90°$ and a skew angle $\beta = 45°$ turned out to work best. The slot length is $L_{SL} = 10mm$, the slot width is $d_{SL} = 0.3mm$. The maximum blowing velocity is $v_{max} \approx 75m/s$, which is about five times the freestream velocity.

The small slot width in combination with high blowing velocities consti- tutes very challenging boundary conditions for the direct numerical simula- tions. In order to resolve all occuring flow scales a very fine grid, especially in wall-normal direction, must be used. The main parameters are set as close as possible to the experiment and are finally: $L_{SL} = 10mm$, $d_{SL} = 2mm$, $v_{max} = 40m/s$.

Although not all main parameters exactly match, qualitative agreement between experiment and numerical simulation is quite encouraging. To illus- trate this, streamwise mean velocity contours and cross-flow velocity vectors in three successive cross sections downstream of the slot are plotted in Figs. 4 and 5. The strong blowing acts like a large obstacle and downstream of the slot actuator a vortex forms transporting high-momentum fluid to near-wall regions. This vortex is persistent and still increases near-wall momentum far downstream of the actuator. This mechanism is clearly observable in experi- mental data and qualitatively reproduced by the numerical simulation.

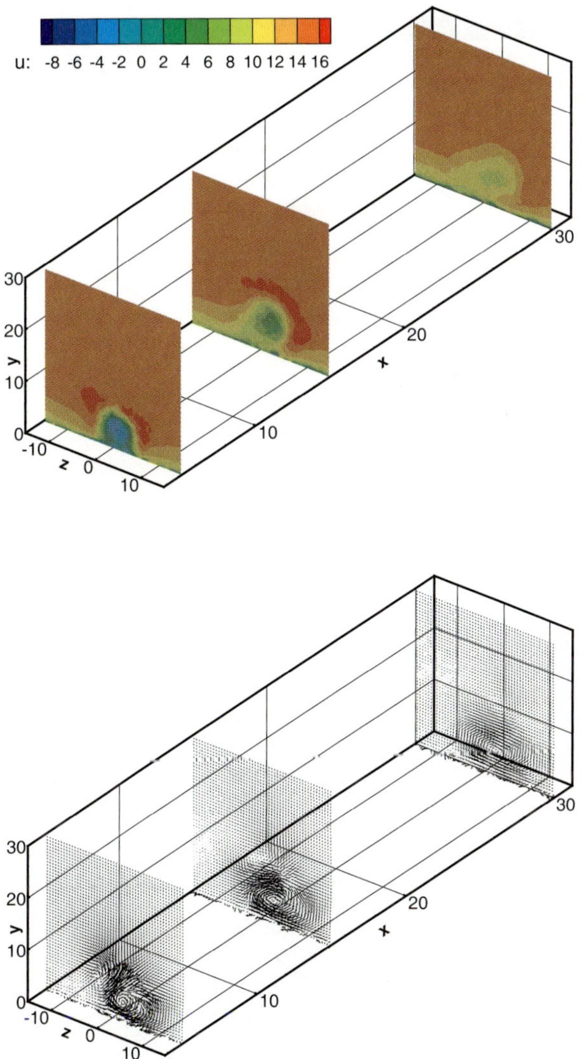

Fig. 4. Contour lines of mean streamwise velocity u (in m/s) (top) and v-w-vectors (bottom) at three planes across main flow direction. Data from experiments provided by University of Braunschweig [3]

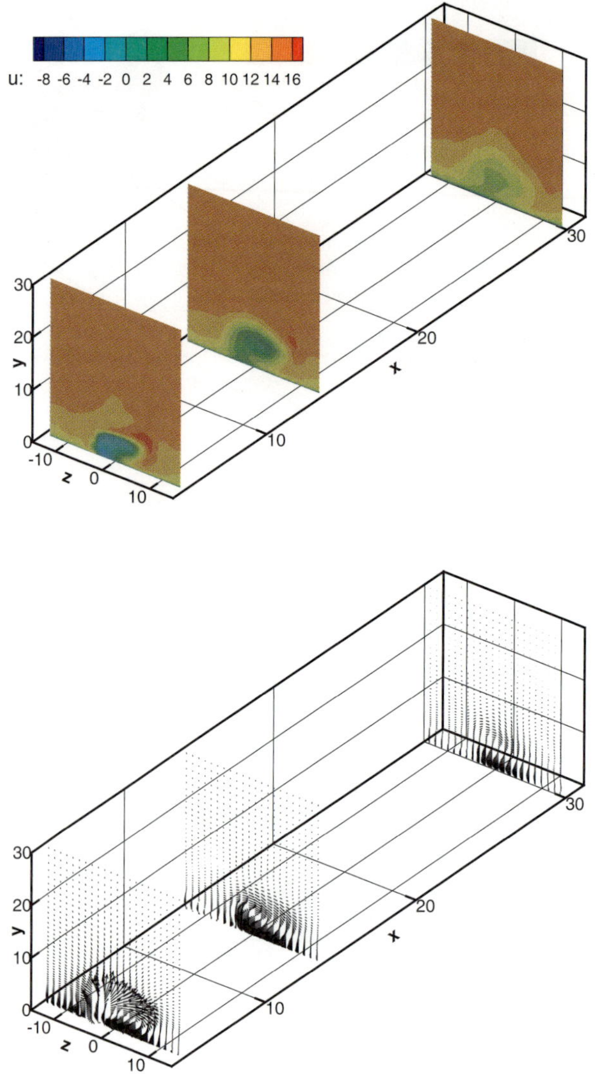

Fig. 5. Contour lines of mean streamwise velocity u (in m/s) (top) and v-w-vectors (bottom) at three planes across main flow direction. Data from unsteady direct numerical simulations

3 Conclusions

A comparison between experiment and unsteady direct numerical simulation of a slot actuator with steady blowing for turbulent boundary-layer separation control has been presented. During experimental studies it turned out that high blowing velocities are required to increase momentum near the wall, and therefore to hinder the flow to seperate from the flow surface in adverse-pressure gradient flows. The numerical effort to simulate such control devices is tremendous, at least with the actual formulation of our DNS solver. Improvements are planned or already under way, like the use of a strechted grid in spanwise direction or domain decomposition with a refined grid in the vicinity of the actuator.

4 TeraFlop Workbench Tunings

Within the TeraFlop Workbench project tunings were introduced to the code to enable new research.

4.1 Introduction

The application was reworked in several stages to provide better scalability and better single CPU performance. The improvements are to enable better utilization of a large machine. Traditionally the program has been run on one maybe two SX nodes. The high performance of the SX-6 machines in comparison to their contemporary competitors made the SX the platform of choice. As single CPU systems or even SMP machines have reached the limits of what is possible in performance the direction is to run constellations of fast machines. This introduces new problems as domain decomposition.

Scaling becomes very important and even if a problem is dividable most problems are not easily parallelized. As the machine is very large with 72 nodes, one wants to take advantage of the power of these nodes to calculate larger and more detailed cases. The application scales with the dimensions of the dataset, this means that distributing the code on less powerful processors, increases the need for a larger domain, which in turn increases the need for computing power. The SX-8 nodes in themselves already are very powerful and by simply scaling the problem from the old usage of one to two nodes, to ten nodes improves throughput and research abilites of the IAG.

4.2 Tunings

Sine and Cosine Transforms

The application is relying on sine and cosinus transforms, since the Z dimension is represented in frequency space. These transforms were compiled from source. The sine transform uses sines only as a complete set of functions in the interval from 0 to 2π, and, the cosine transform uses cosines only. By contrast, the normal Fast Fourier Transform (FFT) uses both sines and cosines, but only half as many of each. Sine and cosine transforms are not a part of the highly optimized mathematical libraires, however FFTs are. By combining sine and cosine transform data, it is possible to use the FFT to do the transform.

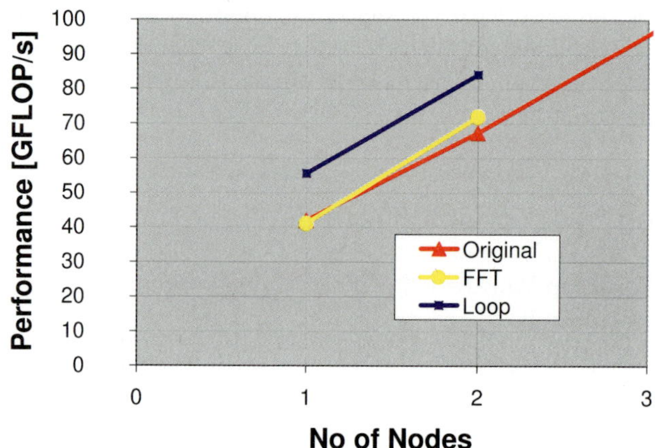

Fig. 6. Improvment by switching to FFT is not so visible in the performance plot as the number of operations went down with the execution time. Single CPU loop improvements also give the performance a boost

Improving Communication

The application does have a rather costly all-to-all communication, when dealing with the frequency domain Z direction of the dataset. One target was to overlap the communication, which can only be made by one CPU of the eight CPU nodes, with meaningfull work, increasing the throughput. Another step is to use the SX-8 global memory option in MPI communication.

The application has the data domain decomposed in the Z direction. In the Z decomposition the Poisson equation is calculated, in an iterative manner using penta-diagonal solvers. Loops in the solvers and loops in general in the program are well vectorized. As the FFT transform needs to be solved a redistribution of the data has to take place. The FFT needs to have access to all values in the Z direction. The way it is made is using a new simple domain decomposition over the X dimension. In the X decomposition the spanwise direction is calculated using sine and cosine transforms. The transforms are implemented by FFTs using the NEC MathKeisan FFT library.

To change between these two representations, all data has to be redistributed between all MPI processes. In the earlier version the communication dealt with the full dataset at once. Instead of doing MPI communication and FFT of the whole data, first MPI communication is done on a Y layer. The communication works in blocks of data "decomposed" in the Y direction. By breaking down also the Y direction in independant blocks, see Fig. 7, it is possible to create a pipelined loop that deals with the different stages. This allows that one SMP thread can deal with the MPI communication, while the other SMP threads can do calculations in the other Y blocks. Thus effectively overlapping the communication done by one CPU with FFT calculation that is done with remaining CPUs. As the first layer is calculated, the next layer is transported. The first layer that already has been communicated can then be calculated on by the FFT algorithm and so on.

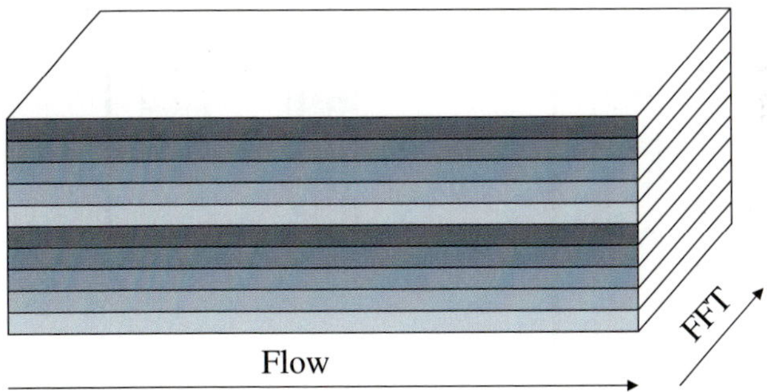

Fig. 7. Treating the Y dimensions as independant blocks in the communication and calculation of the sine-cosine transforms

The stages that need to be considered are

1. Reorganizing the data per CPU
2. Sending the data with `MPI_ALLTOALL`
3. Reorganizing the data for the FFT
   Calculating the FFT
   Reorganizing the data per CPU
4. Sending the data with `MPI_ALLTOALL`
5. Reorganizing the data into the original format

and each block will go through these stages.

The main target behind this new pipelined approach is to overlap the communication with work. As can be seen in Fig. 9, as the single threaded MPI call can be done in parallell, the rest of the CPUs do not have to sit idle. This advantage is only possible in a hybrid program MPI/SMP. As each MPI process can use up to 8 CPUs per node it means that up to 7 CPUs sit idle in the old version in some of the stages (2 and 4).

This work distribution improves scaling as the communication can overlap with the calculation. This is especially important as all data has to be redistributed between all MPI processes at every iteration.

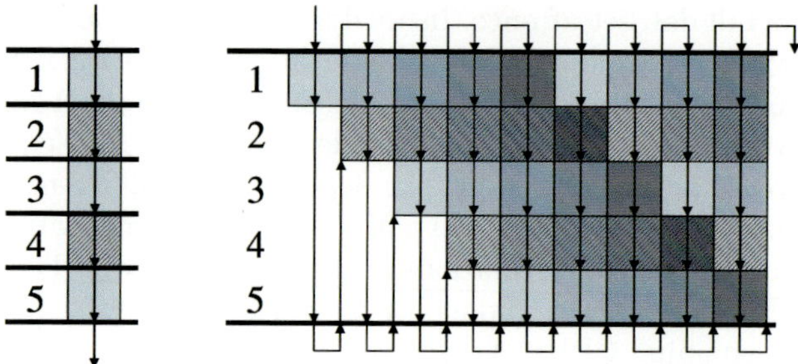

Fig. 8. Working through the blocks in a pipelined manner instead of like before the whole data at once gives less syncronization points between the threads

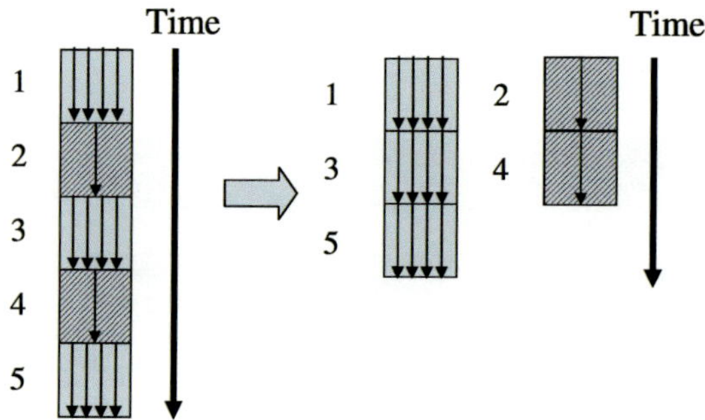

Fig. 9. Overlapping computation with communicaton makes the most out of the computer using the single threaded MPI implementation

Global Memory

The transport bufferes used in the `MPI_ALLTOALL` were placed in the global memory region with compiler directives. Global memory is normaly used by the MPI implementation, and by placing the data there from the program directly databuffer copy can be saved. Global memory regions are still memory local to the machine, therefore not inhibiting performance on the local level.

Asynchronous Communication

As the routines had been adapted for this division in Y, the step of changing the MPI communication was taken. A communication pattern using `MPI_Put` was introduced. This gives each thread the possibility to exchange data to global buffers. This more asynchronous communication can also be achieved on other platforms implementing the `MPI_ALLOC_MEM` call to allocate global memory. The `MPI_ALLOC_MEM` call on the SX allocates global memory, a memory region that is specially treated by the MPI implementation.

5 Results

The tuning steps brought different improvements, single cpu performance and scaling speedups.

The initial change, moving from homemade sine and cosinus transforms gave a direct improvement of 34% for the Z symmetric path and 22% for the Z non-symmetric path. As can be seen in Fig. 6 the performance change is not as visible as the change in realtime as the number of operations also goes down using the library FFT. The larger improvement in the symmetric path can be attributed to the symmetry in the frequency data. Less data is used to reach the result, giving the symmetric path an advantage over the non-symmetric path. Both versions are used in research depending on the problems investigated.

In the scaling step most concentration was put into the non-symmetric path that is the more complex setup. Also the tridiagonal solver was improved with some directives.

The current version shows good performance and scalability for large problems on the SX-8 system. While earlier research was made with smaller models and less nodes (1-2), todays research is demanding larger models. The current models in use have between 90 M and 996 M grid points. This will continue to grow in the future. The primary target is to scale the X dimension to enable studies of different flow phenomena over larger distances.

Figure 10 shows the performance differencs between the different stages. The original performance on 30 nodes for the 314 M grid points dataset was 745 GFLOP/s. With all tunings it reaches 1.4 TFLOP/s on 30 nodes,i.e. it became almost twice as fast. The largest performance step is taken by the overlapping of communication with calculations (47 %), but also the use

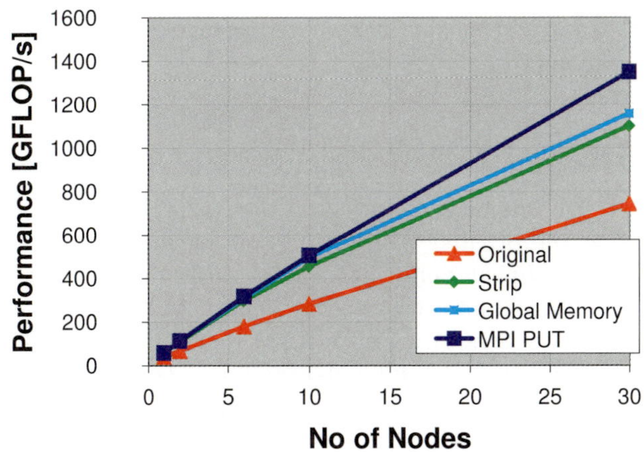

Fig. 10. Sustained performance for a case with 314 M grid points, detailing the different tuning stages

of the asynchronous communication improves scaling (16 %). The usage of global memory does only limit some memory copy but that is always helpful (5 %). The single node performance (only SMP without MPI), reaches 59 GFLOP/s and an efficiency of 46 %. Still the small case on 30 nodes retains an efficiency of 35 %. Less than 10 % drop off when compared to the first MPI measurement using 2 nodes at 44 %. Efficiency is calculated from a 16 GFLOP/s peak performance of one CPU.

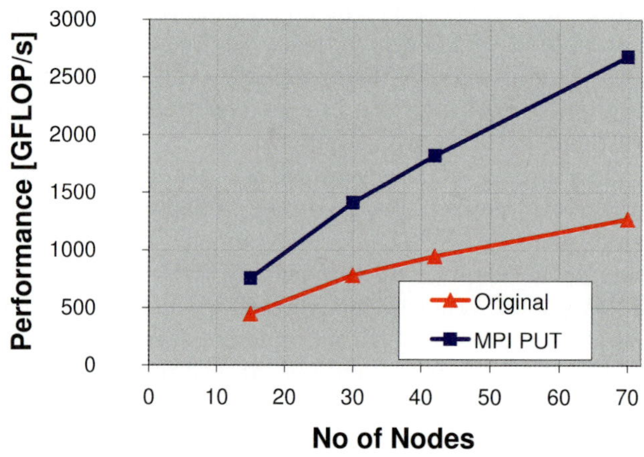

Fig. 11. Sustained performance for a case with 1100 M grid points, detailing the different tuning stages

As can be seen in the Fig. 11, the sustained performance on 70 nodes of NEC SX-8 at HLRS reaches 2.68 TFLOP/s using a large test case with 1100 M grid points. The strong scaling plot between 15 and 70 nodes, show an efficiency between 39 % and 30 %. 1.27 TFLOP/s was the performance of the original program using 70 nodes.

5.1 Computational Results from IAG

The numerical code has been executed on the NEC-SX 8 of the hww GmbH, Stuttgart. Using the original version the code attains 4.1 GFLOP/s of 16 GFLOP/s theoretical peak performance on a single processor at a vector operation ratio of 99% and an average vector length of 222. In runs on 14 nodes the code reaches 407 GFLOP/s with a RAM requirement of 193 GByte. The computation time is $0.9\mu s$ per time step and grid point on a computational grid of $2226{\times}793{\times}224$ (streamwise $\times$ wall-normal $\times$ spanwise) grid points.

References

1. Bonfigli, G., Kloker, M.: Spatial Navier-Stokes simulation of crossflow-induced transition in a 3-d boundary layer. In: Nitsche, W., Heinemann, H.-J.; Hilbig, R. (eds) New Results in Numerical and Experimental Fluid Dynamics II. Proc. 11. AG STAB/DGLR Symposium, NNFM 72. Vieweg Verlag, Braunschweig (1999)
2. Messing, R.: Direkte numerische Simulationen zur diskreten Absaugung in einer dreidimensionalen Grenzschichtströmung. PhD Thesis, Institut für Aero- und Gasdynamik, University of Stuttgart (2004)
3. Scholz, P.: Private Communication, Institut für Strömungsmechanik, TU Braunschweig (2006).
4. Wassermann, P., Kloker, M.: Mechanisms and passive control of crossflow-vortex induced transition in a three-dimemsional boundary layer. J. Fluid Mech., **456**, 49–84 (2002)

Modelling of Reentry Nonequilibrium Flows

M. Fertig[1] and M. Auweter-Kurtz[1]

Universität Stuttgart, Institut für Raumfahrtsysteme, Pfaffenwaldring 31, 70550 Stuttgart, Germany `fertig|auweter@irs.uni-stuttgart.de`

In order to numerically simulate the loads arising during reentry of a space vehicle the nonequilibrium Navier-Stokes URANUS (**U**pwind **R**elaxation **A**lgorithm for **N**onequilibrium **F**lows of the **U**niversity of **S**tuttgart) has been being developed in collaboration of the Institute of Space Systems (IRS) and the High Performance Computing Center Stuttgart (HLRS). URANUS accounts for the complex thermochemical relaxation processes employing sophisticated models for chemistry–energy coupling and gas–surface interactions. This paper will briefly describe the current modeling of the 3D Parallel-Multiblock URANUS code as well as enhanced models which have been tested within the 2D/axisymmetric version of the code.

1 Introduction

For the development of reusable space transport systems, a detailed prediction of the thermal loads during re-entry is essential. For this purpose the URANUS (**U**pwind **R**elaxation **A**lgorithm for **N**onequilibrium **F**lows of the **U**niversity of **S**tuttgart) code for hypersonic non-equilibrium flows has been developed at the Institute of Space Systems IRS of the University of Stuttgart in cooperation with the HLRS within SFB 259[1]. Due to the hypersonic speed of a reentry vehicle a strong shock forms in front of the vehicle. At high altitudes, the drop of the mach number across the bow shock causes an increase in temperature of translational motion of the gas particles to several 10000 K. Due to the low density, chemical nonequilibrium as well as thermal nonequilibrium between translation and internal excitation of electronic, vibrational and rotational degrees of freedom arises. Atoms and ions form in the post shock relaxation area. As a consequence, gas temperature around a vehicle returning from low earth orbit drops to about 6000 K. In order to

---

[1] Sonderforschungsbereich 259, Collaborative Research Center 259: "High Temperature Problems of Reusable Space Transportation Systems"

accurately describe the thermochemical relaxation, sophisticated models for the coupling of chemistry and internal degrees of freedom have been developed and implemented. In addition to the gas–phase relaxation processes the interaction between highly reactive gases and the surface material of a reentry vehicle have to be simulated. Different gas-surface interaction models, radiation exchange models and a heat conduction model for the thermal protection system (TPS) have been developed.

During the development of the 3D Parallel-Multiblock URANUS code the modeling was restricted to simplified models in order to allow for an efficient determination of complex reentry vehicles returning from a low earth orbit (LEO). For the complex X-38 reentry vehicle 22 hours of computation were necessary to solve the Navier-Stokes equations on 1.02 million cells with an average performance of 9.4 GFLOPS on 6 NEC-SX5 processors. Therefore, the currently available code does not require teraflop computing. In order to improve the code the more accurate models from the 2D code will be transferred to the 3D Parallel-Multiblock code. In addition, in order to address entries into the atmospheres of other celestial bodies like Mars, chemistry models for different atmospheric composition have to be implemented. A significant increase of required operations by orders of magnitude will result.

2 URANUS Code

In the nonequilibrium Navier-Stokes URANUS code the governing equations in finite volume formulation are solved fully coupled. [18] The Navier-Stokes equations

$$\frac{\partial \mathbf{Q}}{\partial t} + \frac{\partial (\mathbf{E} - \mathbf{E}_v)}{\partial x} + \frac{\partial (\mathbf{F} - \mathbf{F}_v)}{\partial y} + \frac{\partial (\mathbf{G} - \mathbf{G}_v)}{\partial z} = \mathbf{S} , \tag{1}$$

where

$$\mathbf{Q} = (\rho_i, \rho u, \rho v, \rho w, \rho e_{tot}, \rho_k e_{vib,k})^T \quad (i = 1, ..., n_s, k = 1, ..., n_v) \tag{2}$$

is the conservation vector, ρ_i is the partial density of the species, ρ is the density, u, v and w are the velocity components in x-, y- and z-direction and e is the energy. The conservation vector consist of n_s species continuity equations, where n_s is five for the 3D code accounting for N_2, O_2, NO, N and O. In the 2D code n_s is ten accounting for N_2^+, O_2^+, NO^+, N^+ and O^+ in addition. Assuming charge neutrality, no separate continuity equation is solved for the electrons. Two or three momentum equations are solved in the 2D and the 3D codes, respectively. In addition to the total energy equation three vibrational energy equations for the molecular species N_2, O_2 and NO are solved. While in the 3D code thermal equilibrium between translational and rotational energy is assumed, the 2D code employs an extra energy equation for the rotational excitation of the molecules. Moreover, a separate energy equation if solved for the translational energy of the electrons. $\mathbf{E}$, $\mathbf{F}$, $\mathbf{G}$ are the inviscid flux vectors in x-, y- and z-direction; $\mathbf{E}_v$, $\mathbf{F}_v$, $\mathbf{G}_v$ are the viscous flux vectors. $\mathbf{S}$ denotes the source terms vector for chemical reactions and energy exchange.

2.1 Numerical Scheme

To calculate the steady state solution of the Finite Volume Navier-Stokes equations

$$V\frac{\partial \mathbf{Q}}{\partial t} = \mathbf{R}(\mathbf{Q}) \tag{3}$$

with the conservation variables $\mathbf{Q}$, the volumes V and the sum of the inviscid and viscous fluxes and the source terms $\mathbf{R}(\mathbf{Q})$, the implicit Euler time differencing with the usual Taylor series linearization for $\mathbf{R}$ is applied. The resulting linear system

$$\left(V\frac{1}{\Delta t} - \frac{\partial \mathbf{R}}{\partial \mathbf{Q}}\right) \Delta \mathbf{Q} = \mathbf{R}(\mathbf{Q}) \tag{4}$$

with $\mathbf{Q}^{n+1} = \mathbf{Q}^n + \Delta \mathbf{Q}$ has to be solved for each time step. For $\Delta t \to \infty$ the scheme is exactly Newton's method. Most of the advanced complex surface boundary conditions as well as all other boundary conditions are implemented implicitly. For Newton's method it is not necessary to compute the Jacobian exactly. Approximations can be made to reduce memory requirements. For this reason, the Jacobian is computed with first order inviscid fluxes and a thin shear layer approximation for the viscous fluxes. Further reduction of the memory requirement can be achieved by storing the Jacobian using single precision while computing and storing the fluxes and source terms using double precision. Hence, the memory requirement of the Jacobian is halved. The linear system is solved with the Jacobi-Line-Relaxation-Method. After multiplying with the inverse of the main block-diagonal, two (2D) or four (3D) block-diagonals of the matrix corresponding to the other grid directions are transferred to the right hand side of the linear system. The resulting tridiagonal system is solved by vectorized LU-decomposition. Thereafter, the step is repeated with the other grid directions. This procedure is iterated until convergence is achieved. The linear systems are solved on each block separately. After this solving step the resulting data at the domain boundaries are exchanged in case of a multiblock simulation and another solving step follows. To accelerate the code, the Krylov subspace methods GMRES, BiCGstab, CGS, TFQMR and QMRCGstab with a vectorizable ILU-preconditioner have been implemented. [28]

2.2 Parallelization

The parallel multiblock version of 3D URANUS code is able to deal with nearly any kind of structured multiblock mesh which enables one to simulate re-entry vehicles with complex geometry such as X-38 with body-flap. [1] In a multiblock mesh, beside the surface boundary, physical boundaries (inflow, outflow, symmetry) may occur at each of the six block sides. This limitation is due to the fact that the complex flux-based surface model requires a second order discretization in the transformed computational space. Hence, the implementation is very time-consuming for all six meshblock sides. Furthermore, each block can have neighboring blocks at each block side. When using GridPro meshes, there is exactly one block connected to one block side as a neighbor.

Other multiblock meshes may have more than one neighbor at one block side. The URANUS code is able to handle data exchange with several neighbors at one block side. Moreover, it is able to handle any combination of neighborhood, e.g. a block can have a boundary to one and the same side of a neighbor block at two of its sides. Such layouts exist in current multiblock meshes. The data exchange in P-MB URANUS is performed by MPI. Hence, the portability of the code on widely used supercomputers is guaranteed. In order to maintain second order accuracy in space discretization, the data exchange occurs at the block boundary over an intersection zone (domain overlapping) of two cells. Different blocks may have different local coordinate directions. This has to be considered when exchanging data between two block neighbors. The necessary data conversion is done during the communication and is hidden from the application. To obtain the full performance of the system used, the load balancing tool JOSTLE [31] was implemented into the URANUS code. Applying load balancing, several cases are possible when using multiblock meshes for flow calculations: Large blocks are cut into several pieces, depending on their size, such that the resulting new blocks can be handled as separate blocks. In contrast, blocks which contain only a small number of volumes do not fully utilize a CPU and can be computed together on one processor. The attached load balancing tool is able to find a good distribution of the blocks to the available processors.

2.3 Gas–Phase Modeling

The discretization of the inviscid fluxes of the governing equations is performed in the physical space by a Godunov-type upwind scheme employing Gas Kinetic Flux Vector Splitting (KVFS). [3] In the 2D code a more accurate but less robust approximate Riemann solver from Roe and Abgrall is available in addition. [30] Second order accuracy is achieved by employing van Leer's TVD limited extrapolation. [23] In the 2D code a TVD limited extrapolation of van Albada, an ENO limiter (Essentially Non-Oscillatory) as well as WENO (Weighted Essentially Non-Oscillatory) limiters can be employed as an option. [23, 29] The viscous fluxes are discretized in the transformed computational space by central differences on structured grids using formulas of second order accuracy.

Thermochemical Relaxation

Thermochemical relaxation processes in the gas-phase are accounted for by the advanced multiple temperature Coupled Vibration-Chemistry-Vibration (CVCV) model. [24] The CVCV-model was developed by picking up the concepts of Treanor and Marrone [26, 34] and by extending their dissociation CVDV modeling to exchange and associative ionisation reactions. Later on, the influence of rotational energy was included into the model. [24, 25]

The model is based on state selective reaction rates. The assumption is made that the vibrational energy contributing to overcome the activation barrier is limited by the parameter α to a certain fraction of the activation energy αA. This assumption was made to assure that a minimum fraction of the activation energy comes from the translational energy of the reactants. Therefore, two state selective reaction rates are given which are different in the second exponential term.

By summing up the state selective reaction rates over all rotational-vibrational levels weighted by a Boltzmann distribution function, an analytical expression for the overall reaction rates can be obtained. In addition to these rates the average vibrational and rotational energies gained or removed in chemical reactions are consistently modelled from the state selective rates. Therefore, both the influence of vibrational energy on reaction rates and the influence of chemical reactions on the average internal energy content of the molecules are taken into account. This is a major advantage over other reaction rate models which treat this influence inconsistently, or even neglect this effect. For further details about the CVCV-model see references [24, 25].

Transport Coefficients

In dissociated re-entry flows strong gradients are observed in densities, temperatures and velocities. To describe the exchange of mass, momentum and energy under these conditions, Chapman-Cowling's approximations for the transport coefficients were implemented. [13]

The Chapman-Cowling method was developed from rigorous kinetic gas theory and relates the transport coefficients to the pair potential energy functions of the particles in the gas. [21] The potential energy functions are used to determine the so-called collision integrals. The transport coefficients of monoatomic gases are then approximated by an infinite series in terms of the collision integrals. The number of elements of this series which are taken into account determines the approximation level. It was shown by Devoto [8] that especially in case of ionized gases the accuracy of the second approximation of the heat conductivity is low. In the fully ionized limit of hydrogen Devoto computed heat conductivities differing by 57% between the second and the third approximation. He concluded that at least the next higher level of approximation has to be used for ionized gases to obtain the correct transport coefficients. Since the computation of the higher approximations for a multicomponent mixture is extremely laborious, Deveto separated the Boltzmann equations for the heavy particles and for the electrons. [9] Therefore, the second approximation of the thermal conductivity of the heavy particles was implemented. The third approximation is employed for the thermal conductivity of the electrons in the 2D code. Due to the relatively small difference in species mass, thermal diffusion is neglected up to now. As a side effect, the required computational work for the determination of the thermal conductivity is significantly lowered. For viscosity as well as for the multicomponent diffusion the first approximations are used. In the 2D code the diffusion fluxes are determined employing the Stefan-Maxwell relations. [13] Since only little differences arise when comparing results for dissociated air from Stefan-Maxwell relations with those from Fick's Law for mixtures, only the latter is implemented in the 3D code. [13] Since Fick's Law for mixtures does not conserve mass, mass conservation is guaranteed making use of the flux correction proposed by Sutton and Gnoffo. [33]

The collision integrals from Gupta et al. [19] do not differ much from those given by Yos. [35] Hence, the collision integrals have been updated with those published by Capitelli et al. [2] Since Capitelli's publication contains only interactions with neutral particles, the shielded coulomb collision integrals from Mason et al. [27] are employed for charged species interactions.

2.4 Surface Modeling

Energy exchange and reactions of the gas components at a thermal protection system's surface influences the surface heat flux significantly especially at high altitudes. Therefore, kinetic gas-surface interaction models are implemented in order to take into account near-surface nonequilibrium effects in leeward flows, base flows and general re-entry flows at high altitudes. [5, 7, 11]

Slip Boundary Conditions

At altitudes above approximately 85 km the temperature and velocity jumps at the surface have to be considered for vehicles with a characteristic dimension of $L = 1\,\mathrm{m}$ in order to determine surface loads. For simple chemistry and accommodation models it is possible to determine slip velocity and slip temperature analytically, such that Dirichlet or Van Neumann boundary conditions can be applied [20]. Daiß [5] was the first, who solved flux balance equations at the surface boundary. He showed that it is not necessary to evaluate the slip values analytically. For explanation, a virtual volume element shown in Fig. 1 with edge lengthes Δs and Δn is introduced at the surface. For $\Delta n \rightarrow 0$ the volume of the virtual surface element becomes zero and the numerical scheme given by Eq. 4 reduces to

$$- \left(\partial \mathbf{R} / \partial \mathbf{Q} \right) \Delta Q = \mathbf{R}(\mathbf{Q}). \tag{5}$$

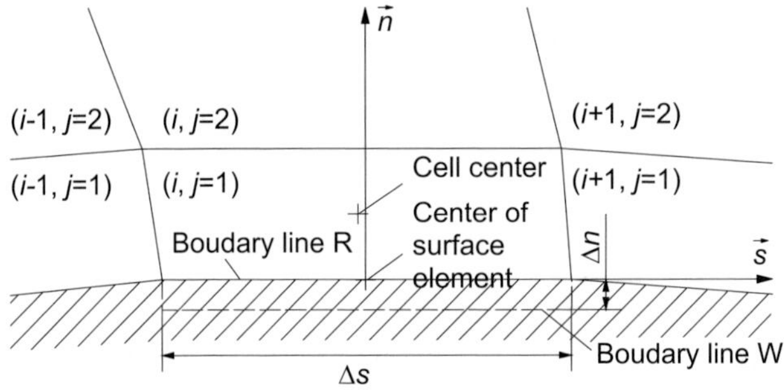

Fig. 1. Grid geometry at the surface with virtual surface volume constituted by the edges Δs and Δn. The cell indices are given in paranthesis. [11]

The residual vector $\mathbf{R}$ then only contains the fluxes accross the boundary lines 'R' and 'W' indicated in Fig. 1, leading to

$$(\partial\mathbf{Q}/\partial t) + (\partial\mathbf{F}/\partial\mathbf{n}) = \mathbf{0}, \tag{6}$$

where $\mathbf{n}$ is the normal vector. Eq. 6 is dicretized by

$$\mathbf{R}\,(\mathbf{Q}) = (\mathbf{F}_{n,W} - \mathbf{F}_{n,R})\,\Delta s. \tag{7}$$

The fluxes at the gas-phase interface $\mathbf{F}_{n,R}$ are computed with the continuum approach described in the previous sections, the flux at the surface interface W is split into fluxes due to particles approaching the surface $\mathbf{F}_n^{(-)}$ and fluxes $\mathbf{F}_n^{(+)}$ due to leaving particles, which both will be shown in the following sections.

The fluxes to the surface $F_{n,i}^{(-)}$ are determined by evaluation of

$$F_{n,i}^{(-)} = \int_{c_{n,i}<0} \Psi_i\, c_{n,i}\, f_i\, d\mathbf{c_i} \tag{8}$$

in the molecular velocity range $-\infty \le c_{n,i} \le 0$ for the transported quantities

$$\Psi_i \in \left\{ m_i,\, m_i\mathbf{c}_i,\, \frac{1}{2}m_i c_i^2,\, m_i h_{0,i},\, \varepsilon_{vib,j}(v),\, \varepsilon_{rot,k}(J) \right\} \tag{9}$$

employing a perturbed Maxwellian partition function as an approximation of the velocity distribution function f_i. [11] In the URANUS code the transported properties are species mass m_i, momentum $m_i\mathbf{c}_i$, kinetic energy $\frac{1}{2}m_i\mathbf{c}_i^2$, chemical energy $m_i h_{0,i}$ with the mass specific formation enthalpy $h_{0,i}$ at $0\,\mathrm{K}$, rotational energy $\varepsilon_{rot,i}(J)$ and vibrational energy $\varepsilon_{vib,i}(v)$. [11]

Note that no flux balance equation is solved for the normal momentum perpendicular to the surface for several reasons. First, a flux equation for normal momentum has a strongly reflecting character, such that flow simulation becomes unstable. Second, the determination of normal momentum is required for all species impinging at the surface and the particles generated in chemical reactions. Third, the pressure gradient perpendicular to the surface is small such that a continuum based boundary equation can be applied. Therefore, the derivative of the momentum equation normal to the surface

$$R_{nn,p} = \nabla^T \cdot \left[\rho \, \mathbf{v} \, \mathbf{v}^T + pE - \eta K \right] \cdot \mathbf{n} \tag{10}$$

is used, where p is the pressure, E is the unit matrix, η is the coefficient of viscosity and K the rate of the shear stress tensor.

The flux vector $\mathbf{F}_n^{(+)}$ is determined dependend on $\mathbf{F}_n^{(-)}$ and numerous gas–surface interaction models which will be briefly described in the following sections.

Non-Reactive Scattering

Energy and momentum of particles change due to collision with the surface. In case of completely diffuse reflection, temperature accommodates to the surface temperature and the mean momentum tangential to the surface of particles leaving the surface becomes zero. Specular reflection of particles on the other hand exchanges neither energy nor tangential momentum with the surface. Only the velocity normal to the surface is inverted. For this kind of non-reactive scattering the simple Maxwell model is implemented in the URANUS code. [4, 20] Accommodation coefficients can may be independently specified for translational, vibrational and rotational energy of the scattered species. This model is sufficiently accurate in flow regimes, where the Navier-Stokes equations in the gas–phase are valid and surface slip applies. At low altitudes the temperature jump between gas and surface vanishes. As a consequence, the influence of energy and momentum accommodation at the surface is negligible.

Catalysis

Chemical reactions of the gas species at the surface may have a significant influence on the heat flux to the surface. At around 70 km of altitude the number of particle collisions in the post shock relaxation zone are sufficiently high to allow for conditions close to equilibrium at the edge of the boundary layer. For a return from low earth orbit nearly complete dissociation of Oxygen and a significant dissociation degree of Nitrogen follow. Roughly $\frac{2}{3}$ of the energy content of the gas are transferred into the chemical degrees of freedom, i.e. formation enthalpy of the atoms.

In comparison to the gas temperature the surface temperature is very low. The temperature of frequently used reusable TPS materials does not exceed 2500 K. As the temperature jump between gas and surface tends to vanish with decreasing altitude, this is also a maesure for the gas temperature close to the surface. At about 2500 K Oxygen and Nitrogen molecules are hardly dissociated. Therefore, the chemical equilibrium condition of the air close to the surface is a mixture of N_2 and O_2. Due to the low pressure and particle density in the boundary layer the number of particle collision does not allow for a complete recombination of the atoms formed in the post shock relaxation zone at high altitudes. Hence, a significant amount of the impinging particles are atomic species. In case of recombination of the atoms at the surface, the reaction enthalpy is released which tends to increase the surface heat flux. For design purposes, non- and fully-catalytic cases can be simulated by URANUS. In addition, advanced models have been developed to account for the finite catalytic behavior of TPS materials. If only recombination is taken into account, three different mechanisms may be distinguished.

- First, the collision probability of the particles with the surface is much higher then the collision probability with other gaseous particles. Due to the higher collision probability, the recombination rate at the surface is also much higher as compared with the gas–phase. Nevertheless, the recombination probability is about 10^{-7} typically, such that no significant influence on surface heat flux arises.
- Second, atoms may become adsorbed, i.e. be trapped at the surface for some time. Since the density of an adsorbed layer is typically of the order of a fluid. Therefore, the collision probability of a gas atom with an adsorbed atom is high. Depending of the number density of adsorbed atoms and the activation energy the recombination probability can become very high. This recombination process is called Eley-Rideal mechanism.
- Third, depending on the surface material adsorbed atoms may move along the surface. Collision probability of such an atom with an immobile one is very high. The recombination of a mobile adsorbed atom with an immobile one is called Langmuir-Hinshelwood mechanism.

Detailed catalysis models accounting for the three different recombination processes under consideration of thermal nonequilibrium effects have been implemented in the 2D code. [6, 10, 16] In addition, starting from state selective rates, the mean energy of reacting particles and of particles formed are determined consistently. However, up to now no significant effect of incomplete chemical accommodation on TPS thermal loads was found. [12]

Due to the different competing recombination mechanisms, the temperature and pressure dependency of the recombination probability is rather complicated. If the pressure dependency is neglected, the recombination probability depends on temperature only. A simplified model was developed, where recombination probability of each species is expressed by a fifth order polynomial. [14] For the determination of the energy of reacting particles complete energy accommodation was assumed. The catalytic behavior of the implemented technical surfaces (SiC, SiO_2) is fitted to the recombination coefficients which were measured by Stewart in a large surface temperature range. [32] Up to now, only these simplified models have been implemented in the multiblock 3D code.

Active and Passive Oxidation of SiC

In the high temperature areas of re-entry vehicles the surface temperature may exceed 2000 K. Therefore, TPS materials based on SiO_2 such as RCG (Reaction Cured Glass) can not be used. Ceramics based on SiC withstand much higher temperatures and have a high emissivity as well. This allows for an effective radiation cooling of the surface. As compared to SiO_2-based materials the catalycity of SiC concerning oxygen and nitrogen atoms is significantly higher at high temperatures. [32] Furthermore, SiC may react with oxygen or nitrogen forming the gaseous species SiO, SiN, CO and CN. If the surface temperature is sufficiently low and the oxygen partial pressure is sufficiently high, a solid SiO_2-layer may form at the surface, which acts as a protection layer for the underlying SiC. All of the reactions described so far are exothermal, i.e. chemical energy is transferred towards the surface. Therefore, a protective SiO_2-layer is desirable at the surface since SiO_2 not only protects the SiC from further oxidation but is also less catalytic. Ambient conditions leading to the formation of a protective SiO_2-layer are called 'passive'. Unfortunately, the protective SiO_2-layer is removed from the surface in the temperature range of 1600 K-2100 K depending on oxygen partial pressure. As a consequence, the bare SiC is exposed to the highly reactive, partially dissociated gas flow. In this case, the reaction behavior is called 'active'.

Depending on the ambient conditions, 'passive oxidation' may be associated with a material gain while 'active oxidation' always leads to a material loss. Under re-entry conditions both passive as well as active oxidation usually lead to a material loss since the gaseous reaction products are transported away from the surface. Therefore, stationary conditions arise if the production of the protective layer and its recession are equal.

In order to describe the transition from active to passive and vice versa three reaction zones shown in Fig. 2 are distinguished. The first reaction zone (RZ_1) describes the interface between SiC and gas, the second one (RZ_2) models the SiO_2–gas interface and the last one (RZ_3) describes the SiO_2–SiC interface and the reactions inside the SiO_2-layer. The thickness of the

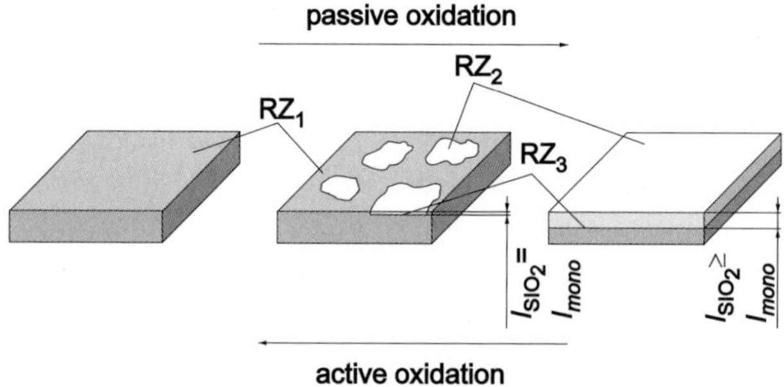

Fig. 2. Schematic illustration of the transition from active to passive oxidation on SiC and vice versa. [16]

SiO_2-layer is determined by stationary conditions of thermal decomposition of SiO_2 and the formation of SiO_2 inside RZ_3. If the surface is completely covered with SiO_2, the heat flux towards the surface equals that of the SiO_2 model accounting for catalytic recombination only. [10]

It is planned to extend the currently available chemical models for the surface with respect to ablative TPS materials which have to be used for high-speed reentry vehicle due to the extremely high surface heat loads. Subsequently, the implementation of the detailed models in the parallel multiblock 3D code is planned.

2.5 Radiative Energy Exchange at the Surface

Most modern TPS materials like SiC or SiO_2 based materials are cooled by radiation according to Stefan-Boltzmann's law

$$q_{i,rad}^{+} = \varepsilon_i(T_{W,i})\, \sigma\, (T_{W,i})^4 \ , \tag{11}$$

where $T_{W,i}$ is the surface temperature, σ is the Stefan-Boltzmann's constant $(5,6697 \cdot 10^{-8}\frac{W}{m^2 K^4})$ and ε_i is the temperature dependent emissivity. Emissivity for SiO_2 and SiC has been fitted in the temperature range between 300K and 2200K. [22] If the surface is concave, a part of the radiative energy flux $q_{i,rad}^{+}$ emitted by a surface element S_i reaches other surface elements. Hence, radiation emitted by a surface S_1 in the direction of S_2 leads to an increase of the surface temperature $T_{W,2}$ and vice versa. In the 3D code, this is taken into account by view factors for flat, lambertian surfaces. Assuming that the reflected part of the radiation does not strike any other surface element, the determination of the view factors was reduced to a purely geometrical problem. [17]

2.6 Heat Conduction in the TPS

Due to the flow during reentry the heat loads and surface temperature strongly differ in different surface areas. In order to reduce TPS weight and cost, the heat shield design is optimized with respect to the expected local heat loads. Therefore, seven different materials are used for the TPS surface of the Shuttle orbiter for example. Generally, high loads arise at the stagnation point and at the wing leading edges. Very high local heat loads arise especially at edges of flaps and rudders but also at the junction of different surface materials, due to the jump in chemical behavior. Typically, a local maximum in heat load is associated with strong gradients in surface temperature in surface lateral direction. Due to heat conduction in the surface material, the surface temperature is significantly lower as compared with a radiation equilibrium assumption. [15,22] In order to cover this behavior, a the 2D URANUS scheme was loosely coupled with the commercial ANSYS finite element solver. [15] Since several coupling steps are necessary, Infed developed a finite element based model for heat conduction within the TPS which gives a new temperature distribution at the surface taking into account thermal conduction within the TPS. [22] The heat conduction equations are solved for each time step of the flow solver. The resulting TPS heat conduction fluxes are then added to the total surface energy flux balance equation.

3 Performance

Due to the advances in computer technology 2D nonequilibrium flows can computed on usual personal computers. Typical turn around time for a fore body flow simulation on 3100 cells are five hours on a 3.2 GHz Pentium IV even for the advanced modeling described previously. For more complex vehicle geometries the number of required grid cells is higher. However, up to now less than 16000 cells were sufficient for an accurate determination of the surface properties of all vehicles examined so far. Therefore, turn around times of a few days follow.

In order to resolve the complex geometry of the X-38 reentry vehicle a mesh consisting of 1.02 million cells was necessary. Simulations for X-38 was performed on a NEC SX-5 computer using 6 CPUs and requiring 17.8 GB of memory. The memory requirements of about 17 KB per cell mainly arises in order to store the flux vector Jacobian for the whole computational domain, see Sect. 2.1. Significant overhead arises due to the required domain overlapping such that e.g. the relative memory requirement of the smallest blocks is nearly twice the one of the biggest blocks. The simulations ran with a total performance of 9.4 GFLOPS on 6 NEC-SX5 processors by an average vector operation ratio of 95.4%. The bad performance results from the greatly differing block sizes with short vector legths of the smallest blocks in combination with a high load imbalance as described below. CFL numbers up to 100 have been reached for the computations with a 2nd order flux discretization; higher CFL numbers were possible for 1st order discretization. On average, each iteration took about 45 seconds, where up to 18% of the computatiual time was spent on communication among the processors. The residual dropped six orders of magnitude which is very satisfying for such complex geometry. However, 600 iterations were necessary to reduce the residual by one order of magnitude in the Newton phase of the convergence. It must be mentioned that it took many iterations to place the shock wave such that the Newton phase of convergence could begin. In total, the simulation took about 22 hours of computation.

The sizes of the blocks differ greatly; the smallest blocks have 9200 cells and the largest block contains 260700 cells. Most of the small-sized mesh blocks were necessary to describe the geometry of the body flap deflection of the X-38 vehicle. At the very beginning of the simulations, the large discrepancy in the number of cells of each block caused a large imbalance of the loads among the processors despite using the balancing tool JOSTLE. After several modifications in the way that the input parameter of the balancer became part of the general input parameters of the code, the previous imbalance of 38% could be reduced to a range of 8% to 10%. It was remarkable that even after finding a good distribution of the load on the processors for one run of the simulations, the following, continuing runs could not maintain this distribution for the same case. This problem has been removed by increasing the coarsing/reduction threshold - the level at which graph coarsing ceases. Two main effects were observed: the partitioning is speeded up and it should reduce the amount of data that needs to be migrated at each repartition, since coarsing gives a more global perspective to the partitioning. It was necessary to move away from the default value 20 to higher values up to 200. This should not happen since the standard value already yielded a good partitioning, which could not be maintained for the following runs. We believe that this is not a result of misusing JOSTLE but rather a problem which requires further careful investigations. Subsequently, focus will be on speedup and scaleup.

4 Summary

The complex modeling of physical and chemical properties of reentry nonequilibrium flows has been briefly described. Due to the limitations of computer performance, only a subset of the advanced available models for thermo-physical relaxation and surface chemistry were implemented in the parallel multiblock 3D code. With the more complex models required for the simulation of flows around interplanetary entry vehicles, an increase of computational resources by orders of magnitude is expected. Therefore, the improvement of the models will also require an algorithmic improvement. Especially, load balancing but also vectorization need further enhancement to allow for an efficient simulation of the loads of future entry vehicles.

5 Acknowledgements

This research work has been performed within the Collaborative Research Center "Sonderforschungsbereich 259".
The authors would like to thank the Deutsche Forschungsgemeinschaft (DFG) and European Space Agency ESA for their support. The support by TETRA is also acknowledged.
Furthermore, the authors would like to express their special thanks to the HLRS for the technical support. Th. Bönisch is also acknowledged for his support.

References

1. Th. Bönisch and R. Rühle. Efficient flow simulation with structured multiblock meshes on current supercomputers. In D.R. Emerson, editor, *Parallel Computing in CFD*. ERCOFTAC Bulletin No. 50, 2001.

2. M. Capitelli, C. Gorse, S. Longo, and D. Giordano. Transport coefficients of high-temperature air species. AIAA-Paper 98-2936, 7th Joint Thermophysics and Heat Transfer Conference, Albuquerque, New Mexico, USA, June 1998.

3. S.Y. Chou and D. Baganoff. Kinetic flux vector splitting for the navier stokes equations. *Journal of Computational Physics*, 130:217–230, 1997.

4. A. Daiß, H.-H. Frühauf, and E.W. Messerschmid. A new gas/wall interaction model for air flows in chemical and thermal nonequilibrium. In *Proceedings of the Second European Symposium on Aerothermodynamics for Space Vehicles and Fourth European High-Velocity Database Workshop*, Noordwijk, The Netherlands, 1994. ESTEC.

5. A. Daiß, H.-H. Frühauf, and E.W. Messerschmid. New slip model for the calculation of air flows in chemical and thermal nonequilibrium. In *Proc. of the Second European Symposium on Aerothermodynamics for Space Vehicles and Fourth European High-Velocity Database Workshop*, pages 155–162, Noordwijk, The Netherlands, 1994. ESTEC.

6. A. Daiß, H.-H. Frühauf, and E.W. Messerschmid. Modeling of catalytic reactions on silica surfaces with consideration of slip effects. AIAA-Paper 96-1903, 1996.

7. A. Daiß, H.-H. Frühauf, and E.W. Messerschmid. Modeling of catalytic reactions on silica surfaces with consideration of slip effects. *Journal of Thermophysics and Heat Transfer*, 11(3), July 1997.

8. R.S. Devoto. Transport properties of ionized monatomic gases. *The Physics of Fluids*, 9(6):1230–1240, June 1966.

9. R.S. Devoto. Simplified expressions for the transport properties of ionized monatomic gases. *The Physics of Fluids*, 10(10):2101–2112, October 1967.

10. M. Fertig. *Modellierung reaktiver Prozesse auf Siliziumkarbid-Oberflächen in verdünnten Nichtgleichgewichts-Luftströmungen*. PhD thesis, Universität Stuttgart, Stuttgart, Germany, URN: urn:nbn:de:bsz:93-opus-24683, URL: http://elib.uni-stuttgart.de/opus/volltexte/2005/2468/, 2005. (in German).

11. M. Fertig and M. Auweter-Kurtz. Flux based boundary conditions for Navier-Stokes simulations. In *Proceedings of the Fifth European Symposium on Aerothermodynamics for Space Vehicles*, ESTEC, Noordwijk, The Netherlands, 2004. ESA.

12. M. Fertig and M. Auweter-Kurtz. Influence of chemical accommodation on re-entry heating and plasma wind tunnel experiments. AIAA-Paper 2006-3816, 9th AIAA/ASME Joint Thermophysics and Heat Transfer Conference, San Francisco, California, USA, 2006.

13. M. Fertig, A. Dohr, and H.-H. Frühauf. Transport coefficients for high temperature nonequilibrium air flows. *AIAA Journal of Thermophysics and Heat Transfer*, 15(2):148–156, April 2001.

14. M. Fertig and H.-H. Frühauf. Detailed computation of the aerothermodynamic loads of the mirka capsule. In *Proceedings of the Third European Symposium on Aerothermodynamics for Space Vehicles*, pages 703–710, ESTEC, Noordwijk, The Netherlands, November 1999. ESA.

15. M. Fertig and H.-H. Frühauf. Strömungs-strukturwechselwirkung bei der x-38 klappe (tetra ap 21365). Technical Report IRS-02 P 08, IRS, Universität Stuttgart, Stuttgart, Germany, December 2002. (in German).

16. M. Fertig, H.-H. Frühauf, and M. Auweter-Kurtz. Modelling of reactive processes at sic surfaces in rarefied nonequilibrium airflows. AIAA-Paper 2002-3102, 8th AIAA Joint Thermophysics and Heat Transfer Conference, St. Louis, Missouri, USA, 2002.

17. M. Fertig, F. Infed, F. Olawsky, M. Auweter-Kurtz, and P. Adamidis. *Recent Improvements of the Parallel-Multiblock URANUS 3D Nonequilibrium Code*, pages 293–310. Springer-Verlag, Berlin, Heidelberg, Germany, 2005. ISBN 3-540-22943-4.

18. H.-H. Frühauf, M. Fertig, F. Olawsky, and T. Bönisch. Upwind relaxation algorithm for reentry nonequilibrium flows. In *High Performance Computing in Science and Engineering 99*, pages 365–378. Springer, 2000.

19. R.N. Gupta, K.P. Lee, R.A. Thomson, and J.M. Yos. A review of reaction rates and transport properties for an 11-species air model for chemical and thermal nonequilibrium calculations to 30000 K. Technical Report TM 85820, NASA, 1990.

20. R.N. Gupta, C.D. Scott, and J.N. Moss. Surface-slip equations for multicomponent nonequilibrium air flow. Technical Report TM 85820, NASA, 1985.

21. J.O. Hirschfelder, C.F. Curtiss, and R.B. Bird. *Molecular Theory of Gases and Liquids*. John Wiley & Sons, New York, 1954.

22. F. Infed, F. Olawsky, and M. Auweter-Kurtz. Stationary coupling of 3d hypersonic nonequilibrium flows and tps structure with uranus. *Journal of Spacecrafts and Rockets*, 42(1):9–21, 2005.

23. S. Jonas. *Implizites Godunov-Typ-Verfahren zur voll gekoppelten Berechnung reibungsfreier Hyperschallströmungen im thermo-chemischen Nichtgleichgewicht*. PhD thesis, Institut für Raumfahrtsysteme, Universität Stuttgart, Germany, 1993. (in German).

24. S. Kanne, O. Knab, H.-H. Frühauf, and E.W. Messerschmid. The influence of rotational excitation on vibration-chemistry-vibration-coupling. AIAA-Paper 96-1802, 1996.

25. O. Knab, H.-H. Frühauf, and E.W. Messerschmid. Theory and validation of the physically consistent coupled vibration-chemistry-vibration model. *Journal of Thermophysics and Heat Transfer*, 9(2):219–226, April 1995.

26. P.V. Marrone and C.E. Treanor. Chemical relaxation with preferential dissociation from excited vibrational levels. *The Physics of Fluids*, 6(9):1215–1221, 1963.

27. E.A. Mason, R.J. Munn, and F.J. Smith. Transport coefficients of ionized gases. *The Physics of Fluids*, 10(8):1827–1832, August 1967.

28. F. Olawsky, F. Infed, and M. Auweter-Kurtz. Preconditioned newton-method for computing supersonic and hypersonic nonequilibrium flows. *Journal of Spacecrafts and Rockets*, 41(6):907–914, 2005.

29. T. Rödiger. Analyse limitierter extrapolationsverfahren zur rekonstruktion von hyperschallströmungen im thermochemischen nichtgleichgewicht. Diplomarbeit IRS 04-S-06, Institut für Raumfahrtsysteme, Universität Stuttgart, Germany, 2004. (in German).

30. P.L Roe. Approximate riemann solvers, parameter vectors and difference scheme. *Journal of Computational Physics*, 43:357–372, 1981.

31. J.R. Shewshuk and O. Ghattas. A compiler for parallel finite element methods with domain-decomposed unstructured meshes. In D.E. Keyes and J. Xu, editors, *Scientific and Engineering Computing*, volume 180, pages 445–450. American Mathematical Society, 1994. Proceedings of the 7th International Conference on Domain Decomposition Methods.

32. D.A. Stewart. Determination of surface catalytic efficiency for thermal protection materials – room temperature to their upper use limit. AIAA-Paper 96-1863, 31st Thermophysics Conference, New Orleans, LA, 1996.

33. A. Sutton and P.A. Gnoffo. Multi-component diffusion with application to computational aerothermodynamics. AIAA-Paper 98-2575, 7th AIAA Joint Thermophysics and Heat Transfer Conference, Albuquerque, New Mexico, USA, 1998.

34. C.E. Treanor and P.V. Marrone. Effect of dissociation on the rate of vibrational relaxation. *The Physics of Fluids*, 5(9):1022–1026, 1962.

35. J.M. Yos. Transport properties of nitrogen, hydrogen, oxygen and air to $30000^\circ K$. Technical Report AD-TM-63-7, Research and Advanced Development Division AVCO Corporation, 1963.

A Lattice Boltzmann HPC Application
in Medical Physics

J.Bernsdorf[1], S.E.Harrison[2], S.M.Smith[2], P.V.Lawford[2] and D.R.Hose[2]

[1] CCRLE, NEC Europe Ltd., Rathausallee 10, D-53757 St.Augustin, Germany
   `j.bernsdorf@ccrl-nece.de`
[2] Academic Unit of Medical Physics, University of Sheffield,
   Royal Hallamshire Hospital, Glossop Road, Sheffield, S10 2JF, UK
   `{s.harrison,p.lawford,d.r.hose}@sheffield.ac.uk`

Abstract

Computer simulations play an increasingly important role in the area of Medical Physics, from fundamental research to patient specific treatment planning. One particular application we address in this paper is the simulation of blood flow and clotting, in both synthetic model geometries and domains created from medical images (in particular, magnetic resonance imaging – MRI and computed tomography – CT). Our focus is on the efficient implementation of the lattice Boltzmann method for this type of medical application, particularly the clotting process.

1 Medical Problem

Cardiovascular disease annually claims the lives of approximately 17 million people worldwide [1]. This covers a wide spectrum of pathologies and is often associated with vessel remodelling, i.e. widening (and often thinning of the wall) or lumen reduction (vessel stenosis, see Fig.1). Atherosclerosis is one particular example of such a pathology which results in the formation of deposits (plaque) within the artery wall. Blood flow disturbances occurring as a secondary effect of narrowing of the vessel lumen, may result in the formation of thrombus. Simulating the formation of such flow-related thrombus contributes to a better understanding of the disease process.

Another potential application of modelling is to aid decision making processes during treatment of cardiovascular disease. One example of this is in the treatment of aneurysms. Aneurysms are extreme widenings which can be, if they rupture, life threatening. One method of treatment involves insertion

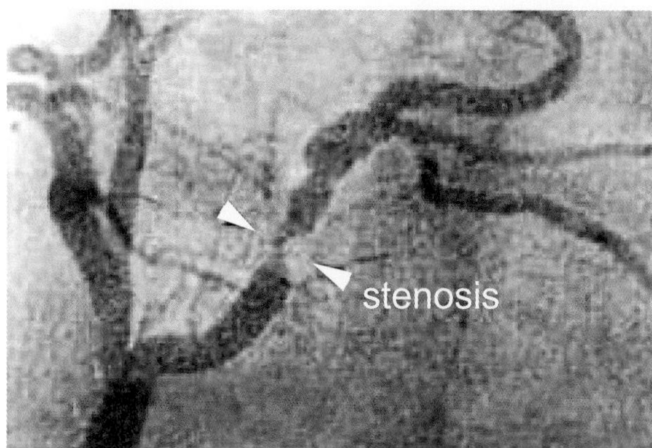

Fig. 1. Angiographic image of a stenosed coronary artery. Diameter of parent vessel is approximately 3mm (image courtesy of Dr. Julian Gunn)

of a metal frame known as a stent, to divert flow from the aneurysm. An alternative is to pack the aneurysm with wire; a procedure known as coiling. The resulting modification of the flow field triggers the process of blood clotting inside the aneurysm and in future, the flow-field following treatment can be predicted by computer simulation. This may ultimately give an insight into the success of the treatment and long-term prognosis. *In vivo* measurements of specific flow properties are possible, but usually not precise enough to predict for example, wall shear stress or pressure distribution with a sufficient spatial resolution. Since invasive treatments of the brain can be problematic, a pre-surgery risk assessment for the likelihood of rupture of the aneurysm in question is a challenging goal.

The use of a stent together with balloon angioplasty is a common method of re-opening a stenosed vessel lumen, and modelling can be used to predict the post-treatment blood flow field. However, the stent material can induce blood clotting and lead to in-stent restenosis, which is an unwanted post-treatment narrowing of the vessel lumen. Specially coated stents can help prevent this effect. In terms of CFD, this is a complex flow, fluid-structure interaction with chemical/biological processes on a variety of timescales.

The role of models such as these are currently being investigated within two European research projects; @neurIST [2] and COAST [3].

2 Numerical Method

The lattice Boltzmann method (LBM) is based on the numerical simulation of a time, space and velocity-discrete Boltzmann-type equation. The propagation and interaction of the particles of an 'artificial computer fluid' are calculated in terms of the time evolution of a density distribution function, representing an ensemble average of the particle distribution. The flow velocity and the fluid density are derived from the moments of the (time and space-discrete) density distribution function, while the pressure is linked to the density by the (model specific) speed of sound. It can be shown theoretically [4] and by a detailed quantitative evaluation of simulation results (for example [5]), that these flow quantities fulfill the time dependent incompressible Navier–Stokes equations under certain conditions.

2.1 Image Segmentation

Discretising the geometry for flow simulations from a CT or MR image is a challenging task. Depending on the applied method, the resulting geometry can vary, and advanced methods must be applied to generate suitable data for the flow simulation. Usually, from these data, a computational grid (or voxel-mesh in the case of lattice Boltzmann, see below) with adequate resolution must be generated.

2.2 Lattice Boltzmann

For simplicity, an equidistant orthogonal lattice is chosen for common LBM computations. On every lattice node $\mathbf{r}_*$, a set of i real numbers, the particle density distributions N_i, is stored. The updating of the lattice essentially consists of two steps (see Eqns. 1,2): a streaming process, where the particle densities are shifted in discrete timesteps t_* through the lattice along the connection lines in direction $\mathbf{c_i}$ to their next neighbouring nodes $\mathbf{r}_* + \mathbf{c_i}$, and a relaxation step, where the new local particle distributions are computed by evaluation of an equivalent to the Boltzmann collision integrals ($\triangle_i^{Boltz}$). For every timestep, all quantities appearing in the Navier–Stokes equations (velocity, density, pressure gradient and viscosity) can be computed locally in terms of moments of the density distribution, and the viscosity is a function of the relaxation parameter ω.

For the present computations, the 3D nineteen–speed (D3Q19) lattice Boltzmann model with single time Bhatnagar–Gross–Krook (BGK) relaxation collision operator $\triangle_i^{Boltz}$ proposed by Qian et al. [6] is used:

$$N_i(t_* + 1, \mathbf{r}_* + \mathbf{c_i}) = N_i(t_*, \mathbf{r}_*) + \triangle_i^{Boltz} \tag{1}$$

$$\triangle_i^{Boltz} = \omega \left(N_i^{eq} - N_i \right) \tag{2}$$

with a local equilibrium distribution function N_i^{eq}:

$$N_i^{eq} = t_p \, \varrho \left\{ 1 + \frac{c_{i\alpha} \, u_\alpha}{c_s^2} + \frac{u_\alpha \, u_\beta}{2c_s^2} \left(\frac{c_{i\alpha} \, c_{i\beta}}{c_s^2} - \delta_{\alpha\beta} \right) \right\} \tag{3}$$

This local equilibrium distribution function N_i^{eq} must be computed every timestep for every node from the components of the local flow velocity u_α and u_β, the fluid density ϱ, a lattice geometry weighting factor t_p and the speed of sound c_s, which we choose in order to recover the incompressible time–dependent Navier–Stokes equations:

$$\partial_t \varrho + \partial_\alpha \left(\varrho u_\alpha \right) = 0 \tag{4}$$

$$\partial_t (\varrho u_\alpha) + \partial_\alpha (\varrho u_\alpha u_\beta) = -\partial_\alpha p + \mu \, \partial_\beta \left(\partial_\beta u_\alpha + \partial_\alpha u_\beta \right) \tag{5}$$

The present LB method is of second–order accuracy in space and time.

2.3 Wall Boundary Conditions

A special feature of the lattice Boltzmann method is the efficient and cheap handling of equidistant Cartesian meshes. In combination with a highly optimised implementation, tens of millions of grid points can be handled on large PCs or workstations. This allows one to use the 'marker and cell' approach for representing the geometry with sufficient accuracy by marking single lattice nodes as occupied or free.

On occupied lattice nodes a so called 'bounce back' wall boundary condition is applied, which simply shifts back the density distributions propagated to the occupied lattice nodes. The inversion of momentum of the shifted distributions leads to the physical zero velocity wall boundary condition.

Arbitrary complex geometries can be approximated by clusters of occupied lattice nodes, similar to a three-dimensional pixel image: the so called 'voxel geometry'. Geometry can either be generated analytically for generic shapes like tubes or a package of spheres or gained from medical imaging techniques such as MRI [7,8].

Changing the geometry during run-time can be achieved by occupying previously free lattice nodes or vice versa without remeshing. An appropriate local rule must be defined which decides at each iteration if a free lattice node turns into a solid or remains unoccupied. Further details of how we exploit this for the clotting simulation will be given in Sect. 2.5.

For an efficient implementation as described in chapter 3, the performance is almost independent of the complexity of the geometry involved. The code used for the simulations presented in this paper is highly optimised for vector–parallel machines such as the NEC SX series, a sustained performance of 25 million lattice site updates per second is achieved on a NEC SX6i (see Sect. 3.2).

2.4 Aging Model

A simple 'aging model' has been suggested [9] to estimate the residence time of previously activated blood. The model requires advection-diffusion simulation of a passive scalar tracer for which the Flekkøy algorithm [10] was used. The local concentration of this tracer indicates the time since activation of the fluid and is required for the clotting model.

2.5 Clotting Model

When the local concentration of the tracer (which is computed at each timestep) reaches a certain threshold, solidification takes place. Within the lattice Boltzmann framework this means a fluid node becomes an obstacle node and the solid surface boundary condition is applied. During subsequent iterations the flow field and age distribution adapt to the new geometry, while further clotting on adjacent fluid nodes may occur.

This allows for the concurrent simulation of solidification and flow, which is believed to be essential for capturing the complex flow related clot morphology.

3 Performance Optimised Implementation

The lattice Boltzmann method is said to be very efficient and easy to implement. But in most cases described in the literature, a simple full matrix implementation is used, where the solid fraction is allocated in the computer memory. Depending on the geometry, this is a considerable waste of resources, and for vector computers, also of CPU-cycles.

3.1 Full Matrix vs. Sparse Implementation

In the framework of a simple full-matrix implementation, the density distribution array for the whole bounding box is allocated in memory. This results in $19 * lx * ly * lz$ REAL numbers for the D3Q19 model for a $lx * ly * lz$ lattice.

Well known methods from sparse matrix linear algebra were first applied to the lattice Boltzmann method by Schultz et al. [11], suggesting storage of the density distribution only for the fluid nodes. This requires keeping an adjacency list for the next neighbours' addresses, but (depending on the geometry) can save considerable memory. Only $N * 19$ REAL numbers for the density distribution (N=number of fluid cells) and $N * 19$ INTEGERs for the adjacency list have to be stored in case of a sparse LB implementation.

For our simulations, a pre-processor was implemented which reads the voxel geometry and generates a list of fluid nodes together with the required adjacency list.

3.2 Performance Measurement

For estimating the efficiency of a full matrix versus a sparse implementation, three figures are of interest:

- **MFLOPS** (million floating point operations per second): in comparison to the theoretical peak performance, this figure indicates how efficient the implementation is for the given hardware.
- **MLUPS** (million lattice site updates per second): the MLUPS give the update rate of the code and thus the total speed of the implementation. High MFLOPS do not necessarily result in high MLUPS, since a more efficient implementation of the equations to be solved (requiring less floating point operations per cycle) could reduce the MFLOPS while actually increasing the MLUPS.
- **MBYTE**: the total memory required to store the density distribution array (and the adjacency list for the sparse code) shows which implementation strategy - sparse or full matrix - is more efficient with regard to memory. Although we expect to require some additional memory for the adjacency list, we save memory from a certain fraction of occupied lattice nodes, since only for the fluid nodes memory has to be allocated.

Geometries

To estimate the performance with respect to the above-mentioned quantities, a set of 12 different geometries was considered; from an empty square box over porous media, packed beds of spheres to physiological geometries like an abdominal aorta and a cerebral aneurysm (see Fig.2).

With a great variety of porosity, specific surface and complexity, these geometries well represent the most typical problem configurations.

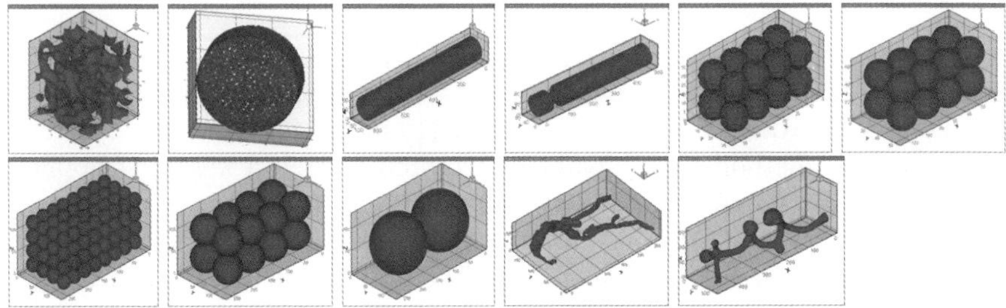

Fig. 2. Benchmark cases 2-12 (increasing numbers from upper left to lower right), case 1 is an empty square channel (not displayed here)

Performance Results

All performance results were achieved on CCRLE's NEC SX6i vector-computer with a peak performance of 8 GFLOPS.

MFLOPS

Figure 3 shows, for all 12 samples, a performance of approximately 4 GFLOPS (50 % of the peak performance) with no strong preference for either method. This indicates that the full matrix and sparse code can be implemented with equivalent performance on a vector-CPU.

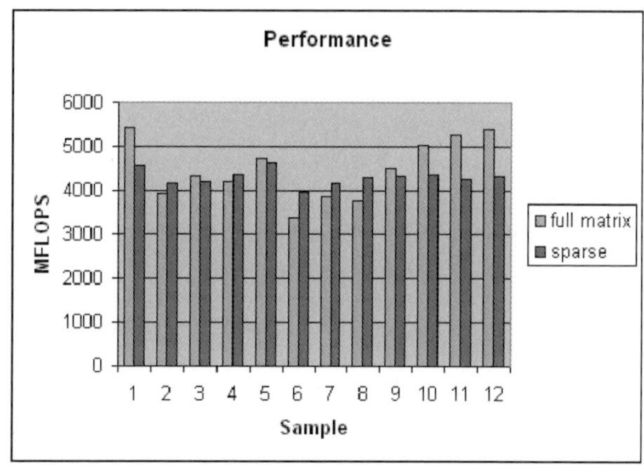

Fig. 3. Full matrix and sparse LB performance (MFLOPS) for the geometries 1-12

MLUPS

A significant performance gap between full matrix and sparse implementation is observed with regard to the lattice site updates: except for the trivial case of a completely empty square box, all MLUPS of the sparse implementation are far above the full matrix case (see Fig.4). This effect is strongest for the medical geometries 11 and 12 (abdominal aorta and cerebral aneurysm) with complex thin channels in a large bounding box, where only a few percent of the domain are fluid nodes: the performance of the full matrix implementation is below 2 MLUPS, while almost geometry-independent, the sparse implementation shows an update rate of approximately 25 MLUPS.

The little additional cost of the indirect address lookup can be seen in case 1 for the square channel: the full matrix implementation using simple index algebra to find the next neighbour cells for the advection step is only $\approx 10\%$ above the sparse code.

MBYTE

The memory required for both methods is of course dependent on the problem size, so large variations can be observed amongst the 12 samples in Fig. 5.

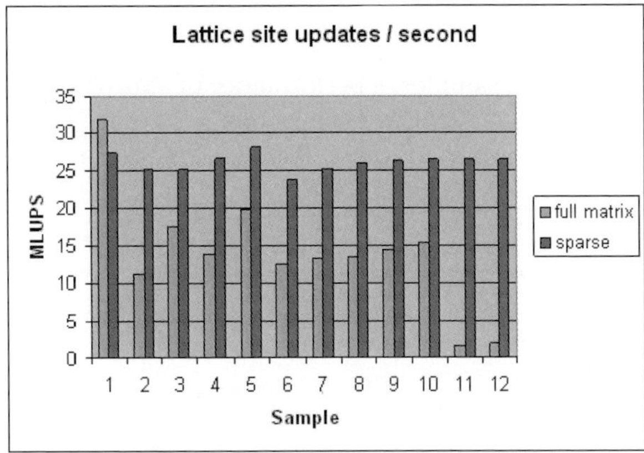

Fig. 4. Full matrix and sparse LB performance (MLUPS) for the geometries 1-12

Except for the free channel (case 1) the memory consumption of the sparse implementation is below that of the full matrix code. The memory reduction, by only allocating fluid nodes, outweighs the cost of storing the adjacency list for all relevant cases. Particularly relevant is the memory reduction to almost 10% for the two medical cases 11 and 12.

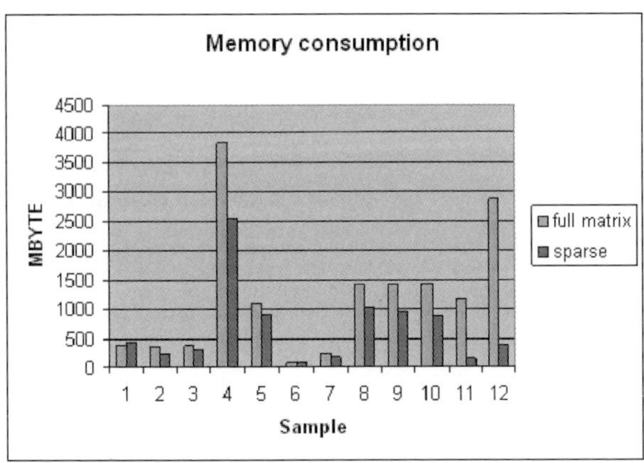

Fig. 5. Full matrix and sparse LB memory consumption (MBYTE) for the geometries 1-12

3.3 Detailed Analysis: Abdominal Aorta

A more detailed analysis for a physiological geometry is carried out for case 11, the abdominal aorta (Fig.6).

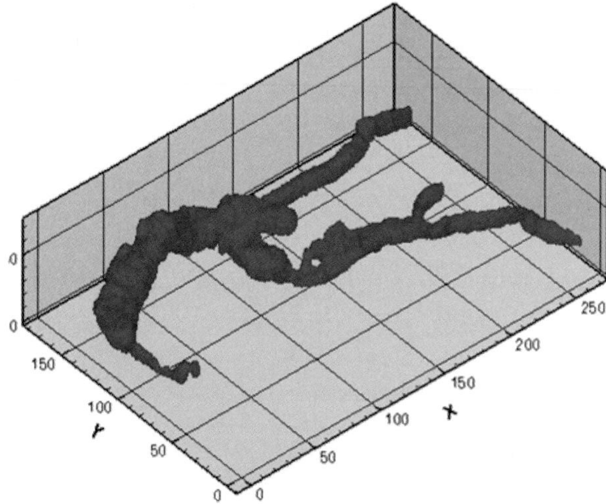

Fig. 6. Geometry case 11: abdominal aorta

The size of the bounding box is $263 * 175 * 74 = 3,405,850$ nodes, of which only 5% (171,166 nodes) are fluid. The figures for both the full matrix and sparse code are listed in Table 1.

Table 1. Comparison of sparse and full matrix performance for a physiological flow case

	Full matrix	Sparse	Comparison (%)
Nodes	3,405,850	171,166	5.0
Memory (MBYTE)	1184	144	12,2
MLUPS	1.6	26.4	1650
CPU seconds	110	6.6	6.0
MFLOPS	5207	4253	81.2

The enormous gain in performance and memory for the sparse implementation is obvious, since the same case can be computed with about 5-6 % of the resources necessary to run a full matrix implementation.

4 Clotting Simulation

For the 2D clotting simulation a lattice size of $lx * ly = 532 * 82$ nodes was used and an initial 200,000 iterations were performed to establish time–dependent flow at $Re = 550$. Following this, the tracer was injected at a constant rate.

Defining a threshold for the tracer concentration, indicating the age of the fluid, allows us to implement the solidification process: all fluid lattice nodes where a concentration above this threshold is found are solidified and no further mass transport is allowed. The threshold concentration was chosen to be small enough to allow clotting within a reasonable simulation time and large enough to avoid solidification of many lattice nodes within a few iterations. Obstruction of the outlet due to clot growing from the walls must also be avoided.

A further 300,000 iterations (equivalent to 11.7 s in real time) were performed to allow a clot to grow (see Fig.7).

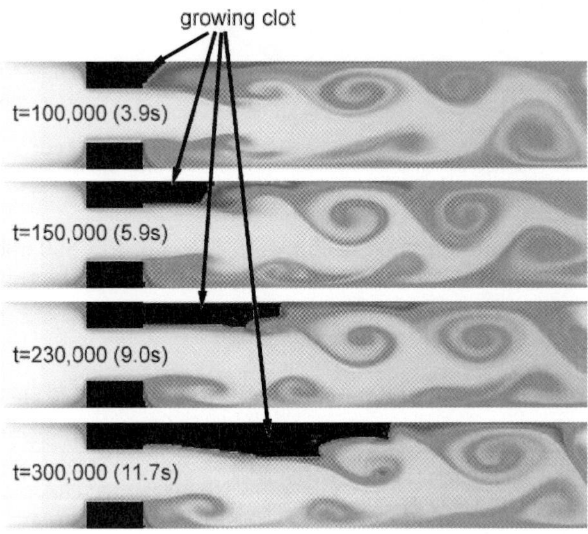

Fig. 7. Growing clot (black) downstream of a 2D stenosis at different time steps t. The age of the fluid is shown in grey, darker regions indicating older fluid

As can be seen in Fig. 7, clot growth initiates in the recirculation domain downstream of the stenosis. The size of the clot increases gradually with time, whilst the flow field adapts to the new geometry. Of particular interest is the downstream migration of the recirculation region. Due to vortex shedding, a secondary vortex is established leading to a second concentration maximum approximately one vortex diameter downstream of the clot (see Fig.8).

A secondary clot has been identified experimentally at this Reynolds number ($Re = 550$), though the relation between this and a secondary vortex must be investigated further.

The final asymmetric shape of the clot reflects the effect of unsteady flow on the pattern formation procedure, showing some qualitative similarities with results of milk clotting experiments using comparable flows.

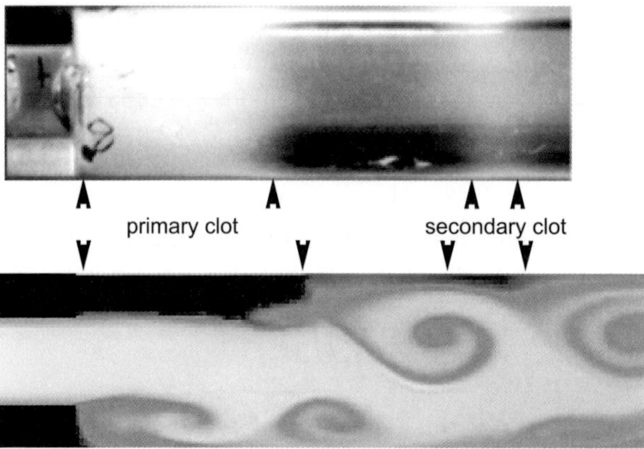

Fig. 8. Secondary milk clot (above) in the experiment and secondary peak in the tracer concentration (below: numerical simulation, darker regions indicating older fluid

A preliminary result of a 3D clotting simulation at $Re = 100$ can be seen in Fig. 9.

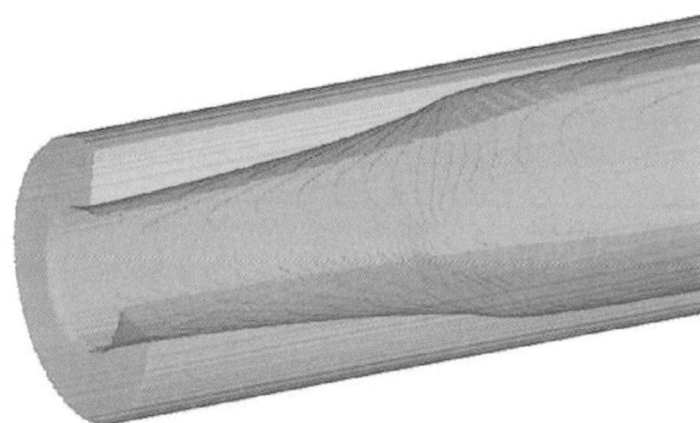

Fig. 9. 3D milk clot downstream of a stenosis at $Re = 100$

5 Conclusion

An equivalent efficient implementation with regard to the peak performance (MFLOPS) can be reached for both the full matrix and the sparse LB implementation, although the sparse algorithm is more complex and requires an indirect address lookup.

A sparse implementation is, for all relevant cases, more efficient with regards to lattice site updates (MLUPS) and memory consumption (MBYTE), which are relevant for the total time and effort a simulation requires. Particularly for the two medical cases, an enormous performance gain and memory reduction was achieved using the sparse implementation.

With a residence-time based clotting model, clotting simulations produce clotting patterns comparable with milk clotting experiments.

Acknowledgements

The lattice Boltzmann flow solver into which the clotting routines were inserted was developed by the International Lattice Boltzmann Software Development Consortium.

References

1. World Health Organisation: Cardiovascular Disease (2006) http://www.who.int/cardiovascular_diseases/resources/atlas/en/
2. "@neurIST: Integrated Biomedical Informatics for the Management of Cerebral Aneurysms" http://www.aneurist.org/
3. "Complex Automata Simulation Technique" http://www.complex-automata.org
4. Frisch U., d'Humières D., Hasslacher B., Lallemand P., Pomeau Y. and Rivet J.-P., Complex Systems **1**, 649-707 (1987)
5. Bernsdorf J., Zeiser Th., Brenner G. and Durst F., Int. J. Mod. Phys. **C**, **9**(8), 1129–1141 (1998)
6. Qian, Y.H., d'Humières D. and Lallemand P., Europhys. Lett., **17**(6), 479-484 (1992)
7. Rothman D.H., Geophysics **53**, 509-518 (1988)
8. Bernsdorf J., Günnewig O., Hamm W. and Münker, O., GIT Labor-Fachzeitschrift **4/99**, 387–390 (1999)
9. Bernsdorf J., Harrison S.E., Smith S.M., Lawford P.V., Hose D.R., icpads, 11th International Conference on Parallel and Distributed Systems - Workshops (IC-PADS'05), 336–340 (2005)
10. Flekkøy E.G., Phys. Rev. E **47**(6), 4247-4257 (1993)
11. Schultz M., Krafczyk M., Tölke J. and Rank E. in: M.Breuer, F.Durst.C.Zenger (Eds.), "High Performance Scientific and Engineering Computing", Lecture Notes in Computational Science and Engineering 21 Springer, Proceedings of the 3rd International Fortwihr Conference on HPSEC, Erlangen, March 12-14, 2001, 114-122, Springer (2002)

Applications II

Molecular Dynamics

Green Chemistry from Supercomputers:
Car-Parrinello Simulations for Ionic Liquids

Barbara Kirchner[1] and Ari P Seitsonen[2]

[1] Lehrstuhl für Theoretische Chemie, Universität Bonn, Wegelerstr. 12, D-53115
Bonn Kirchner@thch.uni-bonn.de
[2] CNRS & Université Pierre at Marie Curie, 4 place Jussieu, case 115, F-75252
Paris Ari.P.Seitsonen@iki.fi

1 Introduction

Ionic liquids have a huge promise in the chemical industry as the solvents for
a wide variety of chemical reactions. Ionic liquids are molten salts, that is,
they consist of anions and cations like the normal food salt NaCl, but unlike
the latter they are liquid already at a low temperature, even down to 100 K.
They carry many useful properties, for example they are non-volatile, *i. e.*
they have a vanishing vapour pressure; thus they can be reused and recycled;
they are in the liquid state in a wide range of temperature 300 K around
the room temperature; they are highly polar; many organic, inorganic and
even polymeric materials are soluble in ionic liquids. Since there are plenty of
possible anions and cations it has been estimate that there are even 10^{16} ionic
liquids. Such a large number enables one to choose a suitable composition for
the wanted process. Examples of cationic and anionic species widely used in
ionic liquids are shown in Fig. 1.

The properties of ionic liquids are dominated by the electro-static interactions between the anions and cations. Figure 2 shows the electro-static potential projected onto an iso-surface of the electron density in one pair of cations
and anions in the ionic liquid $[\text{EMIM}^+][\text{AlCl}_4^-]$. Please notice that there is
hardly any orientational dependence in the potential: Thus the molecules,
even when bonding electro-statically, have hardly any preferential bonding
configuration. This is why the ionic liquid is molten at low temperature, and
the molecules do not escape into vacuum at the surface of the liquid.

Since the ionic liquids are dynamic in nature, a simulation method sampling the combined coordinate-momentum phase space is required in order to
describe the medium satisfactorily. There are two methods which are being
used: The stochastic (probabilistic) Monte Carlo method, in which the atomic
movements are governed by random displacements with selective probabilities,
and molecular dynamics methods, in which the atoms follow Newtonian trajectories deterministically. Both methods provide a series of configurations of

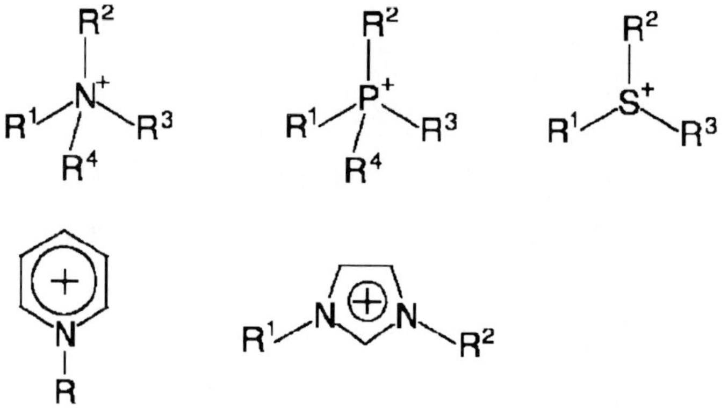

$$BF_4^-, \quad PF_6^-, \quad SbF_6^-, \quad NO_3^-, \quad CF_3SO_3^-,$$
$$(CF_3SO_3)_2N^-, \quad ArSO_3^-, \quad CF_3CO_2^-,$$
$$CH_3CO_2^-, \quad Al_2Cl_7^-$$

Fig. 1. Examples of cations – *top* – and anions – *bottom* – used in ionic liquids

the properties of the material, but only the statistical average over the configurations yields physically meaningful quantities. In the case of molecular dynamics at least 10 ps of simulation time is normally required to describe the basic properties of a liquid well, and since the time step used to integrate the equation of motion is usually $1/2$–1 fs, at least 10^5 evaluations of the forces on the atoms are necessary.

The forces acting on the atoms are often parametrised, that is, they are tabulated for the given interactions needed in a simulation. The parametrisation is done either by fitting to results obtained from experiments or more sophisticated calculations. Therefore the reliability of such simulations is limited to chemical environments similar to the one where the fitting was done, and for example with the most commonly used pair-wise and few-particle interaction models cannot describe chemical bond breaking. An electronic structure method enables one to perform simulations in a wide range of different conditions.

By now there are several studies of ionic liquids using classical force fields, whose accuracy, however, is not guaranteed in such polar molecules and changing relative ordering in ionic liquids. On the contrary, the first simulations including explicitly the electronic structure of a full ionic liquid did not appear until year 2005 [1–3]. They all used DMIM (1,3-dimethyl-imidazolium) as the cation – please see Fig. 3 – but only chlorine ion as the anion. The simulation of a full ionic liquid is much more demanding due to the larger anion needed, and subsequent longer relaxation times. Therefore the employment of supercomputers is compulsory.

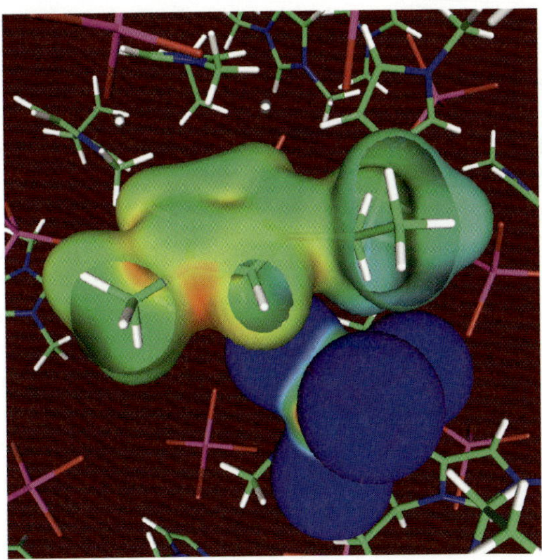

Fig. 2. The iso-surface of the electron density of a single EMIM$^+$ and AlCl$_4^-$ molecules (EMIM = 1,3-ethyl-methyl-imidazolium). The colour scale denotes the electro-static potential: red stands for high and blue for low potential

2 Method

The method that we use to simulate liquids is the Car-Parrinello molecular dynamics. There the atoms are propagated along the Newtonian trajectories, with the forces acting on the ions obtained using the density functional theory solved "on the fly". We shall describe both of the main ingredients of this method in the following. An excellent review is available from D. Marx and Jürg Hutter [4].

2.1 Density Functional Theory

Density functional theory (DFT) [5, 6] is nowadays the most-widely used electronic-structure method. Walter Kohn was awarded the Nobel prise in 1998 for its invention. DFT combines high accuracy in several different chemical environments with reasonable computational effort.

The most frequently applied form of DFT is the Kohn-Sham method. There one solves the set of equations

$$\left\{ -\frac{1}{2}\nabla^2 + V_{KS}\left[n\right]\left(\mathbf{r}\right) \right\} \psi_i\left(\mathbf{r}\right) = \varepsilon_i\psi_i\left(\mathbf{r}\right)$$

$$n\left(\mathbf{r}\right) = \sum_i \left|\psi_i\left(\mathbf{r}\right)\right|^2$$

$$V_{KS}\left[n\right]\left(\mathbf{r}\right) = V_{ext}\left(\{\mathbf{R}_I\}\right) + V_{H}\left(\mathbf{r}\right) + V_{xc}\left[n\right]\left(\mathbf{r}\right)$$

Here $\psi_i\left(\mathbf{r}\right)$ are the Kohn-Sham orbitals, or the wave functions of the electrons; ε_i are the Kohn-Sham eigenvalues, $n\left(\mathbf{r}\right)$ the electron density (can be also seen as the probability of finding an electron at position $\mathbf{r}$) and $V_{\mathrm{KS}}\left[n\right]\left(\mathbf{r}\right)$ is the Kohn-Sham potential, consisting of the attractive interaction with the ions in $V_{\mathrm{ext}}\left(\{\mathbf{R}_I\}\right)$, the electron-electron repulsion $V_{\mathrm{H}}\left(\mathbf{r}\right)$ and the so-called exchange-correlation potential $V_{\mathrm{xc}}\left[n\right]\left(\mathbf{r}\right)$. The latter incorporates the many-body interactions.

The Kohn-Sham equations are in principle *exact*. However, whereas the analytic expression for the exchange term is known, it is not the case for the correlation, and even the exact expression for the exchange is too heavy to be evaluated in practical calculations for large systems. Thus one is forced to rely on approximations. The mostly used one is the generalised gradient approximation, GGA, where one at a given point includes not only the magnitude of the density – like in the local density approximation, LDA – but also its first gradient as an input variable for the approximate exchange correlation functional. The accuracy of the results is often acceptable or even very good, and still the evaluation of the exchange correlation term of the Kohn Sham potential constitutes only a negligible effort.

One further complication in the Kohn-Sham equations is that since the Kohn-Sham potential depends on the electron density and the electron density on the electronic wave function, which are solved *already expecting to know* the Kohn-Sham potential. Thus there is a circular dependence among these variables. In practise this problem is usually solved by iterating the Kohn-Sham equations until self-consistency is reached. This is normally the electronic ground state of the system, or the lowest-energy solution of the electronic structure in the presence of the ionic distribution in $V_{\mathrm{ext}}\left(\{\mathbf{R}_I\}\right)$. Earlier is was the practise to write the Kohn-Sham equations as an eigenvalue problem and solve it by direct diagonalisation of the Hamiltonian matrix expressed in the chosen basis set. However, often the solution of the Kohn-Sham equations is achieved more efficiently with iterative eigenvalue solvers such as conjugate gradient or residual minimisation techniques.

In order to solve the Kohn-Sham equations in a computer they have to be discretised using a basis set. One, a straight-forward choice is to sample the wave functions on a real-space grid at points $\{\mathbf{r}\}$. Another approach, widely used in condensed state systems, is the expansion in the plane wave basis set,

$$\psi_i\left(\mathbf{r}\right) = \sum_{\mathbf{G}} c_i\left(\mathbf{G}\right) e^{i\mathbf{G}\cdot\mathbf{r}}$$

Here $\mathbf{G}$ are the wave vectors, whose possible values are given by the unit cell of the simulation.

One of the advantages of the plane wave basis set is that there is *only one* parameter controlling the quality of the basis set. This is the so-called cut-off energy E_{cut}: All the plane waves within a given radius from the origin are included in the basis,

$$\frac{1}{2} |\mathbf{G}|^2 < E_{\mathrm{cut}} \, ,$$

are included in the basis set. Typical number of plane wave coefficients in practise is of the order of 10^5 per electronic orbital.

The use of plane waves necessitates a reconsideration of the spiked external potential due to the ions, $-Z/r$. The standard solution is to use pseudo potentials instead of these hard, very strongly changing functions around the nuclei [7]. This is a well controlled approximation, and reliable pseudo potentials are available for most of the elements in the periodic table. A more recent and more accurate approximation is the projected augmented wave (PAW) method introduced by Peter Blöchl [8]. However, it is slightly more complicated than the normal pseudo potential method.

When the plane wave expansion of the wave functions is inserted into the Kohn-Sham equations it becomes obvious that some of the terms are most efficiently evaluated in the reciprocal space, whereas other terms are better executed in real space. Thus it is advantageous to use fast Fourier transforms (FFT) to exchange between the two spaces. Because one usually wants to study realistic, three-dimensional models, the FFT in the DFT codes is also three dimensional. This can, however, be considered as three subsequent one-dimensional FFT's with two transpositions between the application of the FFT in the different directions.

The numerical effort of applying a DFT plane wave code mainly consists of BLAS and FFT operations. Thus the methods are well suited for parallel computing, because the previous one generally require quite little communication. However the latter one requires more complicated communication patterns since in larger systems the data needed to perform the FFT on needs to be distributed on the processors. Yet the parallelisation is quite straightforward and can yield an efficient implementation, as recently demonstrated in IBM Blue Gene machines [9]; combined with a suitable grouping of the FFT's one can achieve good scaling up to tens of thousands of processors with the computer code CPMD.

Car-Parrinello Method

The Car-Parrinello Lagrangean reads as

$$\mathcal{L}_{\mathrm{CP}} = \sum_I \frac{1}{2} M_I \dot{\mathbf{R}}_I^2 + \sum_i \frac{1}{2} \mu \left\langle \dot{\psi}_i \middle| \dot{\psi}_i \right\rangle - E_{\mathrm{KS}} + constraints \tag{1}$$

where $\mathbf{R}_I$ is the coordinate of ion I, μ is the fictitious electron mass, the dots denote time derivatives, E_{KS} is the Kohn-Sham total energy of the system and the holonomic constraints keep the Kohn-Sham orbitals orthonormal as required by the Pauli exclusion principle. From the Lagrangean the equations of motions can be derived via Euler-Lagrange equations:

$$M_I \ddot{\mathbf{R}}_I(t) = -\frac{\partial E_{KS}}{\partial \mathbf{R}_I}$$

$$\mu \ddot{\psi}_i = -\frac{\delta E_{KS}}{\delta \langle \psi_i |} + \frac{\delta}{\delta \langle \psi_i |} \{constraints\} \tag{2}$$

The velocity Verlet is an example of an efficient and accurate algorithm widely used to propagate these equations in time.

The electrons can be seen to follow fictitious dynamics in the Car-Parrinello method, *i. e.* they are not propagated in time physically. However, this is generally not needed, since the electronic structure varies much faster than the ionic one, and the ions see only "an average" of the electronic structure. In the Car-Parrinello method the electrons remain close to the Born-Oppenheimer surface, thus providing accurate forces on the ions but simultaneously abolishing the need to solve the electronic structure exactly at the Born-Oppenheimer surface. Several studies have demonstrated the high accuracy and usability of the Car-Parrinello method.

A great advantage of the Car-Parrinello method over the straight-forward Born-Oppenheimer molecular dynamics is that in the former the electrons follow the dynamical trajectory at each step when the electrons and ions and moved simultaneously, whereas in the latter one always has to converge the electronic structure as close to the Born-Oppenheimer as possible. However, because in the latter there is always a remaining deviation from the minimum due to insufficient convergence in the self-consistency, the ionic forces calculated have some error. This leads to a drift in the total conserved energy. On the other hand in the Car-Parrinello method one has to make sure that the electrons and ions do not exchange energy, *i. e.* that they are adiabatically decoupled. Also the time step used to integrate the equations of motion in the Car-Parrinello molecular dynamics has to be 6-10 times shorter than in the Born-Oppenheimer dynamics due to the rapidly oscillating electronic degrees of freedom. In practise the two methods are approximately as fast, and the Car-Parrinello method has a smaller drift in the conserved quantities, but the ionic forces are weakly affected by the small deviation from the Born-Oppenheimer surface.

3 Results so far

So far we have studied the pure $AlCl_3$ [10] and one EMIM molecule in $AlCl_3$ [11] – please see Fig. 3 for an illustration of the species found in the simulations. The bare $AlCl_3$ served us as an initial step to the world of ionic liquids, even if it does not form an ionic liquid alone. In $AlCl_3$ the monomer units bind together, as a monomer unit is energetically unfavourable over $(AlCl_3)_N$ "chains". The simulation cell consisted of 32 or 64 $AlCl_3$ units, leading to 128 or 256 atoms, with a cut-off energy of 25 Ry and average temperature of 540 K, because the experimental melting temperature of $AlCl_3$ is 463 K. The longest simulation run with 64 units was 11.2 ps of production (after equilibration).

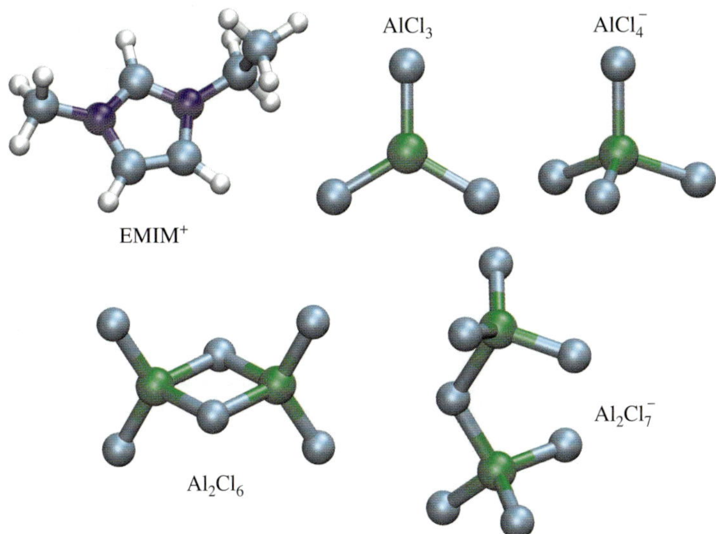

Fig. 3. Cation EMIM = 1,3-ethyl-methyl-imidazolium and neutrals and anions formed by $AlCl_3$ in EMIM-$AlCl_3$ based ionic liquids

The radial distribution functions and angular distribution function in Fig. 4 clearly point towards a edge-sharing bonding pattern (the one shown in Fig. 3) in liquid $AlCl_3$: The distances between the Al-Al and Al-Cl in the former and the Al-Cl-Al angle around $90°$ are characteristic for this conformer, excluding the corner-sharing model. This is changed when the Cl^- counter-ion is added to the liquid $AlCl_3$ with the $EMIM^+$ anion, in which case the $Al_2Cl_7^-$ "opens" into the corner-sharing conformer [11]. This explanation and differentiation of the two conformers of the $Al_2Cl_x^-$ units clarifies the experimentally observed bonding pattern in the $AlCl_3$-based ionic liquids.

Furthermore, using a geometry criterion for bonding between Al and Cl atoms we analysed the formation of chains longer than the dimers in the liquid $AlCl_3$. Figure 5 demonstrates that there are few monomer units, but quite many trimer, tetramer and pentamer units bound in the liquid. The probability of even longer chains is vanishing. In addition we saw dynamic changes in the Al-Cl bonding network: Two initially unbound $AlCl_3$ units come to interact, binding through two shared Cl bridges (Al-Cl-Al). After some time the units might dissociate, but the units take a different Cl atom than initially with them, leading to "Cl exchange" between the units. This is a complicated process involving bond formation and breaking, and would thus be difficult to model using classical, parametrised models.

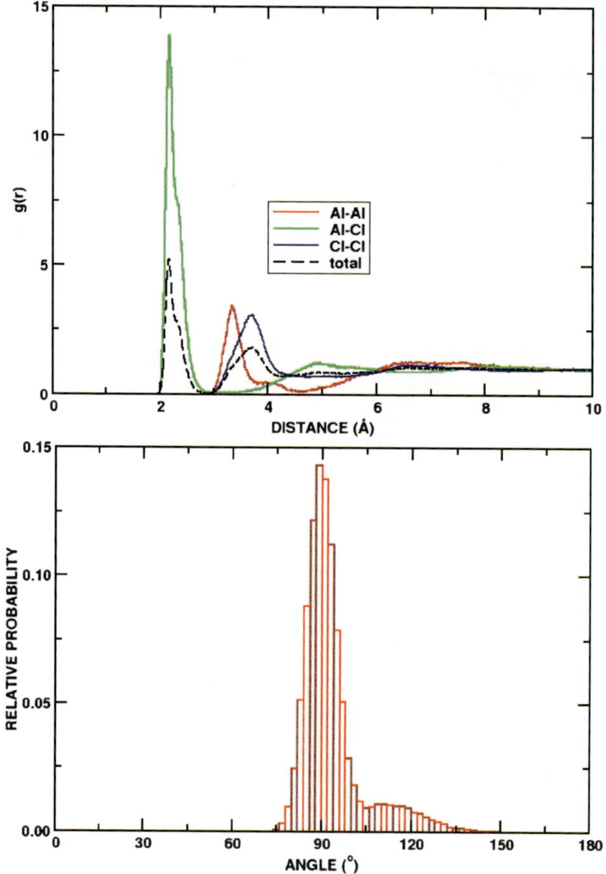

Fig. 4. Partial and total radial distribution functions – *top* – and the Al-Cl-Al angular distribution function – *bottom* – in liquid $AlCl_3$. From Ref. [10]

4 Outlook

We wish to study realistic ionic liquids on the NEC SX-8 installed at HLRS. Massive computing time is not only necessary due to the large number of degrees of freedom, but also because we want to perform *ab initio* molecular dynamics, and at least $1 \ldots 2 \times 10^5$ simulation steps are needed for each single trajectory. This includes the initial, compulsory equilibration period and the rest, about 10 ps of simulation time, can be used for the analysis.

We want to use the code CPMD [12] for these simulations. CPMD has been very efficiently optimised and parallelised. On the NEC SX-8 floating point performance of 13 GFLOPs per processor was achieved in our initial tests on EMIM+$AlCl_3$ systems. The good parallel efficiency is shown in Table 1, where we show the wall-clock time per molecular dynamics step as a function of the

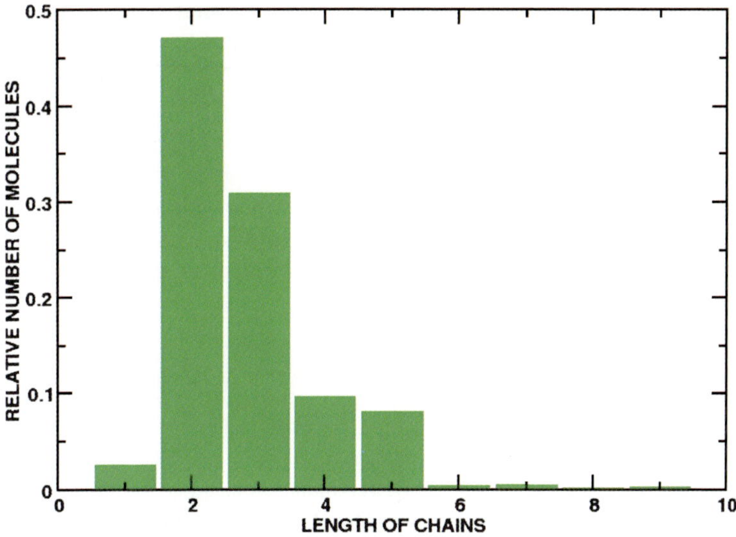

Fig. 5. The distribution of $(AlCl_3)_N$ chain length in liquid $AlCl_3$. From Ref. [10]

number of processors in three different ionic liquid systems. They contain 32, 48 or 64 molecular pairs, *i. e.* altogether 64, 96 or 128 molecules, or 768, 1152 or 1536 atoms.

Table 1. Wall-clock time in seconds per molecular dynamics step of CPMD v3.9.1 on the NEC SX-8 at HLRS in ionic liquid EMIM+$AlCl_3$ with different number of molecular pairs ranging between 32 and 64

		# processors			
		16	32	64	128
	il-64	813	421	224	137
Model	il-48		153	85	
	il-32		52	31	

From the scaling we think that we could justify using 128 processors with 48 molecular pairs for our simulation. We note that the timings were done with version 3.9.1 of CPMD. However, after the release of this version the code has been further optimised on NEC architectures, in particular in the Earth Simulator in Japan.

We want to emphasise that without the possibility to use super-computing resources it is impossible to be able to perform such simulations *at all*. We are convinced that the results achieved from these simulations justify the huge computational cost associated with such a project.

Acknowledgements

We are grateful to Prof. Jürg Hutter for countless valuable discussions and support throughout the project, and to Stefan Haberhauer (NEC) for executing the benchmarks on the NEC SX-8 and optimising CPMD on the vector machines.

References

1. B. L. Bhargava and S. Balasubramanian. Intermolecular structure and dynamics in an ionic liquid: A Car-Parrinello molecular dynamics simulation study of 1,3-dimethylimidazolium chloride. *Chem. Phys. Lett.*, 417, 2005. http://dx.doi.org/10.1016/j.cplett.2005.10.050.
2. Mario G. Del Pópolo, Ruth M. Lynden-Bell, and Jorge Kohanoff. Ab initio molecular dynamics simulation of a room temperature ionic liquid. *J. Phys. Chem. B*, 109, 2005. http://dx.doi.org/10.1021/jp044414g.
3. Michael Bühl, Alain Chaumont, Rachel Schuhammer, and Georges Wipff. Ab initio molecular dynamics of liquid 1,3-dimethylaimidazolium chloride. *J. Phys. Chem. B*, 109, 2005. http://dx.doi.org/10.1021/jp0518299.
4. J. Hutter and D. Marx. Proceeding of the february conference in jülich. In J. Grotendorst, editor, *Modern Methods and algorithms of Quantum chemistry*, page 301, Jülich, 2000. John von Neumann Institute for Computing. http://www.fz-juelich.de/nic-series/Volume1/.
5. P. Hohenberg and W. Kohn. Inhomogeneous electron gas. *Phys. Rev.*, 136:B864 – B871, 1964.
6. W. Kohn and L. J. Sham. Self-consistent equations including exchange and correlation effects. *Phys. Rev.*, 140:A1133 – A1139, 1965.
7. W. E. Pickett. Pseudo potential methods in condensed matter applications. *Comput. Phys. Rep.*, 115, 1989.
8. P. E. Blöchl. Projector augmented-wave method. *Physical Review B*, 50:17953–17979, 1994. http://dx.doi.org/10.1103/PhysRevB.50.17953.
9. Jürg Hutter and Alessandro Curioni. Car-parrinello molecular dynamics on massively parallel computers. *ChemPhysChem*, 6:1788–1793, 2005.
10. Barbara Kirchner, Ari P Seitsonen, and Jürg Hutter. Ionic liquids from car-parrinello simulations, Part I: Liquid $AlCl_3$. *J. Phys. Chem. B*, 110, 2006. http://dx.doi.org/10.1021/jp061365u.
11. Barbara Kirchner and Ari P Seitsonen. Ionic liquids from car-parrinello simulations, Part II: EMIM in Liquid $AlCl_3$. In preparation.
12. CPMD Copyright IBM Corp 1990-2006, Copyright MPI für Festkörperforschung Stuttgart 1997-2001. See also www.cmpd.org.

Molecular Dynamics on NEC Vector Systems

Katharina Benkert[1] and Franz Gähler[2]

[1] High Performance Computing Center Stuttgart (HLRS)
   University of Stuttgart
   70569 Stuttgart, Germany
   benkert@hlrs.de
   http://www.hlrs.de/people/benkert
[2] Institute for Theoretical and Applied Physics (ITAP)
   University of Stuttgart
   70550 Stuttgart, Germany
   gaehler@itap.physik.uni-stuttgart.de
   http://www.itap.physik.uni-stuttgart.de/~gaehler

Summary. Molecular dynamics codes are widely used on scalar architectures where they exhibit good performance and scalability. For vector architectures, special algorithms like Layered Link Cell and Grid Search have been developed. Nevertheless, the performance measured on the NEC SX-8 remains unsatisfactory. The reasons for these performance deficits are studied in this paper.

Keywords: Molecular dynamics, Vector architecture

1 Introduction

The origins of molecular dynamics date back to 1979 when Cundall and Strack [1] developed a numerical method to simulate the movement of a large number of particles. The particles are positioned with certain initial velocities. The relevant forces between the particles are summed up and Newton's equations of motion are integrated in time to determine the change in position and velocity of the particles. This process is iterated until the end of the simulation period is reached. For molecular simulations the particles only interact with nearby neighbors, so usually a cut-off radius delimits the interactions to be considered.

Since this time, the method has gained an important significance in material science. The properties of metals, ceramics, polymers, electronic, magnetic and biological materials can now be studied to understand material properties and to develop new materials. This progress has been made possible by the construction of accurate and reliable interaction potentials for many different kinds of materials, the development of efficient and scalable parallel algorithms, and the enormous increase of hardware performance. It is now possible to simulate multi-million atom samples over time scales of nanoseconds on a routinely basis, an application which clearly belongs to the domain of high performance computing. Such system sizes are indeed required for certain purposes, e.g. for the simulation of crack propagation [2] or the simulation of shock waves [3].

For these and similar applications with high computing requirements, NEC and the High Performance Computing Center Stuttgart (HLRS) formed the Teraflop Workbench [4], a public-private partnership to achieve TFlops sustained performance on the new 72 node SX-8 installation at HLRS.

In this paper, the differences in the implementation of a molecular dynamics program on scalar and vector architectures are explained and an investigation of performance problems on the NEC SX-8 is presented.

2 Implementing Molecular Dynamics Simulations

The dynamics of a system of particles is completely determined by the potential energy function U of the system, shortly denoted as potential. Using Newton's law, the force F_i acting on an atom i is equal to $-\nabla_i U$. These equations are then integrated to retrieve the trajectories of the atoms in course of time. The potential can be simply modeled as an empirical pair potential such as the Lennard-Jones potential, but many systems require more elaborate potential models. For metals, so-called EAM potentials [5,6] are widely used:

$$E = \sum_{i,j} \Phi_{ij}(r_{ij}) + \sum F_i(\rho_i)$$

where

$$\rho_i = \sum_j \Psi_j(r_{ij}).$$

Although being a many-body potential, EAM potentials are as easy to compute as pair potentials.

For short-range interactions, only particles having a distance smaller than the cut-off radius r_c are assumed to contribute to the forces. The algorithmic problem is to construct a parallelizable algorithm scaling linearly with system size to find interacting atom pairs quickly. Testing all possible combinations results in an $O(N^2)$ algorithm, where N is the number of atoms. A first step to reduce the computational effort is the use of Verlet lists [7]: all particles having a distance smaller than $r_c + r_s$, where r_s is the so-called skin, are saved to temporary lists. As long as one of the particles has not moved more than $\frac{r_s}{2}$, those lists can be used for the computation of the potential. To finally obtain an $O(N)$ algorithm, a grid with cells having side lengths slightly greater than $r_c + r_s$ is introduced. At first, the particles are inserted into the cells, and then, in a second step, the Verlet lists are constructed by considering only particles in the surrounding cells, resulting in the Link Cell (LC) method of Quentrec et al. [8] described well in Allen and Tildesley [9]. Parallelization is easily realized using geometric domain decomposition with additional buffer cells [10].

2.1 Implementation on Scalar Architectures

On scalar architectures, the Verlet lists are implemented as two lists, one having pointers into the other list, which in turn contains all particles with distances smaller than $r_c + r_s$. The implementation of the kernel, comprising the calculation and update of the forces, is straightforward. To achieve better cache-usage, all information local to a cell, e.g. particle positions, distances and Verlet lists, can be stored together. Although this introduces an extra level of indirection, execution times decrease.

2.2 Implementation on Vector Architectures

In contrast to scalar architectures, which depend on effective cache usage, vector architectures use pipelining to achieve high performance. Therefore, vector arithmetic instructions operate efficiently on large, independent data sets. Standard molecular dynamics codes are not well suited for vector architectures. Frequent if-clauses, e.g. when deciding whether particles interact or not, and short loop lengths over all particles that interact with a given one prohibit successful vectorization.

For this reason, new algorithms like Layered Link Cell (LLC) [11] and Grid Search (GS) [12] were developed which both use vectorization over all cells instead of vectorization over all particles within one cell. The performance of these algorithms on the NEC SX-8 has been investigated in [13]. Analogously to the LC algorithm, LLC uses cells with side lengths slightly greater than $r_c + r_s$ allowing several particles in one cell. The GS algorithm uses a finer grid with only one particle per cell, which simplifies vectorization, but complicates the choice of an optimal cell length and the distribution of particles into cells. Its runtime compared to LLC is generally lower since more advanced techniques like Neighbor Cell Assignments and Sub-Cell Grouping are used. The Verlet lists are organized as two lists, saving every particle pair whose distance is smaller than $r_c + r_s$.

2.3 The Molecular Dynamics Program IMD

IMD [14] is a software package for classical molecular dynamics simulations developed using C. It supports several types of interactions, like central pair potentials, EAM potentials for metals, Stillinger-Weber and Tersoff potentials for covalent systems, and also more elaborate many-body potentials like MEAM [15] or ADP [16]. A rich choice of simulation options is available: different integrators for the simulation of the various thermodynamic ensembles, options that allow to shear and deform the sample during the simulation, and many more.

Its main design goals were to create a flexible and modular software achieving high performance on contemporary computer architectures, while being as portable as possible. Preprocessor macros allow to switch between scalar and vector versions of the code.

The performance of IMD on several architectures is shown in Table 1. On the SX-8, IMD implements the LLC algorithm. The "mono" option limits calculations to one atom type by hard-coding the atom type as zero. On the SX-8, this gives a considerable performance improvement. In order to allow for maximal flexibility, potentials are specified in the form of tabulated functions. For the pair potential, a Lennard-Jones potential was used. It can clearly be seen, that the price/performance ratio of IMD on vector architectures is dissatisfying.

Table 1. Timings for IMD in μs per step per atom for a sample with 128k atoms

Machine, compiler	pair	EAM
SX-8, mono, sxf90	1.93	2.73
SX-8, sxf90	2.16	3.68
Itanium2, 1.5 GHz, icc	2.58	5.05
Opteron, 2.2 GHz, icc	4.41	6.59
Xeon 64bit, 3.2 GHz, icc	4.64	7.44

3 Performance of the Test Kernel

To better understand the problems of molecular dynamics simulations on the NEC SX-8, a test kernel using the GS algorithm was implemented using Fortran 90.

As test case, a fcc crystal with 131k atoms was simulated for 100 time steps using a calculated Lennard-Jones potential. All following tables show extracts of performance analyses using the flow trace analysis tool *ftrace*. Since the usage of *ftrace* did hardly influence the execution time, statistical profiling results are not included in this paper.

The column "PROG. UNIT" displays the name of the routine or region, "FREQ." gives the number of times a routine was called. "EXCLUSIVE TIME" is the total time spent in the routine and it does not include time spent in other routines called by it. "MFLOPS" depicts the performance in millions of floating point operations per second. The vector operation ratio, i.e. the ratio of vector elements processed to the total number of vector operations, and the average vector length are given in the columns "V.OP RATIO" and "AVER. V.LEN", respectively. These metrics state which portion of the code has been vectorized and to what extent. The average vector length is bounded by the hardware vector length of 256. The time spent waiting until banks recover from memory access is given in the column "BANK CONFLICT".

Table 2 clearly illustrates that nearly all time is spent during force calculation. Although major portions of the force calculation are vectorized and possess a good average vector length of 225.8, the performance of 3.7 GFlops is unsatisfactory.

Update times per step per atom are $0.860\mu s$. As IMD shows only a modest performance difference between tabulated and calculated Lennard-Jones potentials, this can be compared with the results of Table 2, which shows that the Fortran kernel using GS is about twice as fast as IMD using LLC.

Table 2. *Ftrace* performance output for the kernel

PROG. UNIT	FREQ.	EXCLUSIVE TIME[sec](%)	MFLOPS	V.OP RATIO	AVER. V.LEN	BANK CONF
total	113	11.336 (100.0)	3729.1	99.80	225.8	0.1199
forcecalculation	100	11.247 (99.2)	3717.7	99.81	225.8	0.1185

The structure of the kernel is divided into three parts: the construction of the lists of interacting particle pairs and at times the update of the Verlet lists, the calculation of the potential, and the update of the forces.

```
if (verlet lists need to be updated) then
        - find potentially interacting particles
        - build new verlet lists
        - build lists of interacting particles and save distances in
          x, y and z direction as well as squared distance to arrays
else ! old verlet lists are used
        - find interacting particles
        - build lists of interacting particles and save distances in
          x, y and z direction as well as squared distance to arrays
end if
- calculate potential
- update forces
```

3.1 Construction and Use of Verlet Lists

If the Verlet lists need to be updated and there are particles at a given neighbor-cell-relation, the distances between those particles are calculated. If the distance is smaller than $r_c + r_s$, the particles need to be inserted into the Verlet lists. If the distance is also smaller than r_c, the particle numbers as well as the distances are saved to arrays for later use.

The performance characteristics of the construction of the Verlet lists are given in Table 3 and show the same behavior as those of the total kernel: although vectorization ratio and average vector length are good and the number of bank conflicts is small, the performance is low.

Table 3. *Ftrace* performance output for construction of Verlet lists

PROG. UNIT	FREQ.	EXCLUSIVE TIME[sec](%)	MFLOPS	V.OP RATIO	AVER. V.LEN	BANK CONF
newlist	241	0.274 (2.4)	2880.2	99.71	256.0	0.0569

The key problems are the complicated loop structure with nested if-clauses and the high number of copy operations. The frequency with which the Verlet lists need to be updated depends on the skin r_s and on the amount of atomic motion. When simulating a solid, intervals between Verlet list updates are typically 5 – 15 time steps, or even more when simulating at low temperature, whereas for the simulation of liquids more frequent updates may be necessary.

If the old Verlet lists are still valid, the distances between the particles have to be calculated. Those particles which actually interact are stored together with their distances into temporary arrays. The results are shown in Table 4.

Table 4. *Ftrace* performance output when old Verlet lists are used

PROG. UNIT	FREQ.	EXCLUSIVE TIME[sec](%)	MFLOPS	V.OP RATIO	AVER. V.LEN	BANK CONF
oldlist	6930	6.033 (53.2)	3613.5	99.83	225.8	0.0231

The major problems are again the high number of copy operations and the indirect access to retrieve the positions of the particles stored in the Verlet lists.

3.2 Calculation of Potential

As interaction model, a calculated Lennard-Jones potential was used. Given that 16 floating point operations and only two memory operations are needed for one force evaluation, the performance of 9217.4 MFlops is not remarkable (Table 5).

Table 5. *Ftrace* performance for calculation of Lennard-Jones potential

PROG. UNIT	FREQ.	EXCLUSIVE TIME[sec](%)	MFLOPS	V.OP RATIO	AVER. V.LEN	BANK CONF
calcpotential	7171	1.220 (10.8)	9217.4	99.69	225.9	0.0002

Unfortunately, calculated potentials are not often used. For real applications, tabulated potentials fitted to reproduce results from DFT simulations are more flexible, which increases the number of memory accesses and therefore reduces the performance further.

3.3 Update of Forces

During the update of the forces, the distance components in x-, y- and z-direction are multiplied by the calculated force and divided by the distance, and the result is added to the forces of the two particles.

```
do i = 1, nInterAc
  sx(i) = sx(i) * forceOverDistance(i)
  sy(i) = sy(i) * forceOverDistance(i)
  sz(i) = sz(i) * forceOverDistance(i)
end do
```

```
!CDIR NODEP
  do i = 1, nInterAc
    Fx   (interAcList2(i)) = Fx   (interAcList2(i)) + sx(i)
    Fy   (interAcList2(i)) = Fy   (interAcList2(i)) + sy(i)
    Fz   (interAcList2(i)) = Fz   (interAcList2(i)) + sz(i)
    Fxtmp(interAcList1(i)) = Fxtmp(interAcList1(i)) - sx(i)
    Fytmp(interAcList1(i)) = Fytmp(interAcList1(i)) - sy(i)
    Fztmp(interAcList1(i)) = Fztmp(interAcList1(i)) - sz(i)
  end do
```

As can be seen from the above code segment, heavy indirect addressing is needed which is reflected in the performance (Table 6).

Table 6. *Ftrace* performance output for force update

PROG. UNIT	FREQ.	EXCLUSIVE TIME[sec](%)	MFLOPS	V.OP RATIO	AVER. V.LEN	BANK CONF
updateforces	7171	3.669 (32.4)	2121.5	99.82	225.9	0.0378

The update of the forces is the most critical part of the total force calculation. The percentage of time spent in this routine and the low performance due to heavy indirect addressing is a major cause for the unsatisfactory total performance.

4 Summary

Molecular dynamics simulations on vector machines suffer from the latencies involved in indirect memory addressing. For our test kernel using GS, most time is spent when using old Verlet lists and updating forces, whereas simulations with IMD (using LLC) are dominated by the time spent during force updates. Since the reasons for the low performance lie within the structure of LLC and GS, an improvement can only be achieved by developing new algorithms.

5 Acknowledgments

The authors would like to thank Uwe Küster of HLRS as well as Holger Berger and Stefan Haberhauer of 'NEC High Performance Computing Europe' for their continuing support.

References

1. Cundall, P., Strack, O.: A distinct element model for granular assemblies. Geotechnique **29(1)** (1979) 47–65
2. Rösch, F., Rudhart, C., Roth, J., Trebin, H.R., Gumbsch, P.: Dynamic fracture of icosahedral model quasicrystals: A molecular dynamics study. Phys. Rev. B **72** (2005) 014128
3. Roth, J.: ω-phase and solitary waves induced by shock compression of bcc crystals. Phys. Rev. B **72** (2005) 014126
4. http://www.teraflop-workbench.de/.
5. Daw, M.S., Baskes, M.I.: Semiempirical, quantum mechanical calculation of hydrogen embrittlement in metals. Phys. Rev. Lett. **50** (1983) 1285–1288
6. Daw, M.S., Baskes, M.I.: Embedded-atom method: Derivation and application to impurities, surfaces, and other defects in metals. Phys. Rev. B **29** (1984) 6443–6453
7. Verlet, L.: Computer experiments on classical fluids: I. Thermodynamical properties of Lennard-Jones molecules. Phys. Rev. **159** (1967) 98–103
8. Quentrec, B., Brot, C.: New methods for searching for neighbours in molecular dynamics computations. J. Comput. Phys. (1973) 430–432
9. Allen, M., Tildesley, D.: Computer simulation of liquids. Clarendon Press (1987)
10. Plimpton, S.J.: Fast parallel algorithms for short-range molecular dynamics. J. Comput. Phys. **117** (1995) 1–19
11. Grest, G., Dünweg, B., Kremer, K.: Vectorized link cell fortran code for molecular dynamics simulations for a large number of particles. Comp. Phys. Comm. **55** (1989) 269–285
12. Everaers, R., Kremer, K.: A fast grid search algorithm for molecular dynamics simulations with short-range interactions. Comp. Phys. Comm. **81** (1994) 19–55
13. Gähler, F., Benkert, K.: Atomistic simulations on scalar and vector computers. In: Proceedings of the 2nd Teraflop Workshop, HLRS, Germany, Springer (2005)
14. Stadler, J., Mikulla, R., Trebin, H.R.: IMD: A software package for molecular dynamics studies on parallel computers. Int. J. Mod. Phys. C **8** (1997) 1131–1140 http://www.itap.physik.uni-stuttgart.de/~imd.
15. Baskes, M.I.: Modified embedded-atom potentials for cubic materials and impurities. Phys. Rev. B **46** (1992) 2727–2742
16. Mishin, Y., Mehl, M.J., Papaconstantopoulos, D.A.: Phase stability in the Fe-Ni system: Investigation by first-principles calculations and atomistic simulations. Acta Mat. **53** (2005) 4029–4041

Large-Scale *Ab initio* Simulations for Embedded Nanodots

R. Leitsmann, F. Fuchs, J. Furthmüller, and F. Bechstedt

Institut für Festkörpertheorie und -optik
Friedrich-Schiller-Universität Jena
Max-Wien-Platz 1, 07743 Jena, Germany
roman|fuchs|furth|bechsted@ifto.physik.uni-jena.de

Summary. We present the equilibrium interface geometries for (110), (100), and (111) PbTe/CdTe interfaces. The *first principles* calculations are based on large supercells containing a large number of atoms, which have to be treated fully quantum mechanically. The corresponding interface energies are calculated and used to predict the thermodynamic equilibrium crystal shape (ECS) of embedded PbTe nanodots in a CdTe host matrix. These ECSs are used as a starting point for *ab initio* structural optimizations of the embedded PbTe-dots. The results are compared with recent high resolution cross-sectional transmission microscopy investigations.

1 Introduction

Nanostructuring of semiconductors is the modern way of developing devices for electronic, optoelectronic, and sensoric applications. The huge efforts made towards matter manipulation at the nanometer scale have been motivated by the fact that desirable properties can be generated by modifying the spatial quantum confinement of electrons and holes, for instance, by changing the system dimension and shape. Very recently the formation of PbTe quantum dots in a crystalline CdTe host matrix has been demonstrated [1]. High resolution cross-sectional transmission microscopy (HRXTEM) studies for the annealed PbTe/CdTe systems show the existence of rather ideal PbTe nanocrystals with (111), (100), and (110) interfaces with the CdTe host. An intense room-temperature mid-infrared luminescence could be observed at this system. Since the availability of light sources in the mid-infrared spectral region is crucial for many applications, e.g. in molecular spectroscopy and gas-sensor systems for environmental monitoring or medical diagnostics, it is crucial to develop a deeper theoretical understanding of these effects.

2 Computational Method

2.1 Kohn-Sham Energy Functional

To investigate ground-state properties like, e.g. interface structures on an atomic scale, we are applying density functional theory (DFT) [2], in which the ground state energy E_0 of a N electron system in an external potential $v_{ext}(\mathbf{r})$ is given by the solution of the minimization problem of the energy functional $E[n]$ with respect to the electron density $n(\mathbf{r})$:

$$E_0 = E[n]_{n=n_0} = \left[T_s[n] + \int d^3r\, v_{ext}(\mathbf{r})n(\mathbf{r}) + \right.$$

$$\left. + \frac{1}{2} \int\int d^3r\, d^3r'\, \frac{n(\mathbf{r})n(\mathbf{r}')}{|\mathbf{r} - \mathbf{r}'|} + E_{xc}[n] \right]_{n=n_0}. \tag{1}$$

The functionals $T_s[n]$ and $E_{xc}[n]$ give the kinetic energy of a system of N non-interacting electrons and the exchange-correlation energy of a N electron system of ground-state density $n_0(\mathbf{r})$, respectively. The two other contributions to E_0 are the energy of the electrons in the external potential and the Hartree energy.

This many electron problem can be mapped onto a system of non-interacting electrons $\{\psi_i\}_i^N$ that has the same ground state density [3] and that can be represented by a set of one-particle equations, the Kohn-Sham equations:

$$\left\{ -\frac{\hbar^2}{2m}\nabla^2 + v_{ext}(\mathbf{r}) + v_H[n](\mathbf{r}) + v_{xc}[n](\mathbf{r}) \right\} \psi_i(\mathbf{r}) = \varepsilon_i \psi_i(\mathbf{r}), \tag{2}$$

$$n(\mathbf{r}) = \sum_{i=1}^{N} |\psi_i(\mathbf{r})|^2. \tag{3}$$

The terms $v_H[n](\mathbf{r})$ and $v_{xc}[n](\mathbf{r}) = \delta E_{xc}/\delta n(\mathbf{r})$ represent the Hartree and the exchange-correlation potential, respectively. Solving equations (2), (3) self-consistently yields the exact ground state density $n_0(\mathbf{r})$ and thus all physical properties that are functionals of this density.

For a numerical solution we have to expand the wavefunctions and potentials into a certain basis set. For systems with periodic boundary conditions like, e.g. bulk crystalline structures or repeated supercells, an expansion into plane waves

$$\psi_i(\mathbf{r}) = \psi_{n\mathbf{k}}(\mathbf{r}) = \sum_{\mathbf{G}} c_{n\mathbf{k}}(\mathbf{G})e^{i(\mathbf{k}+\mathbf{G})\mathbf{r}}$$

yields the most efficient numerical algorithms. However, representing the rapid oscillations of wavefunctions near the nuclei demands a large number of plane waves. On the other hand in the interstitial region, where the wavefunctions are rather smooth, most of the interesting physical properties are determined. Therefore we employ the Projector Augmented Wave method (PAW) [4] to establish a one-to-one correspondence between the exact, near the nuclei rapidly

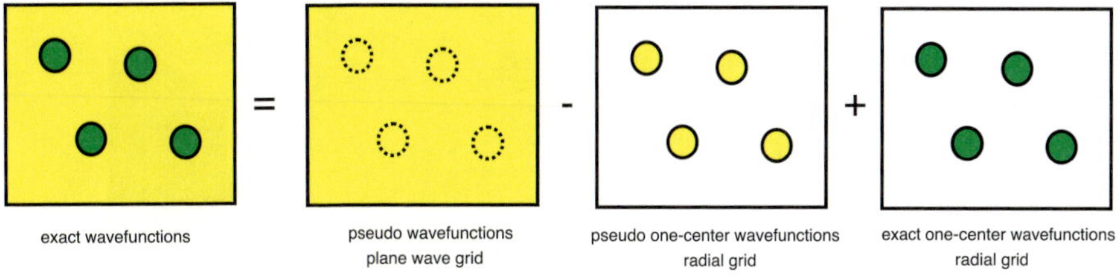

Fig. 1. Schematic picture of the PAW method

oscillating wavefunction $\psi_{n\mathbf{k}}(\mathbf{r})$ and a set of smooth pseudo-wavefunctions $\tilde{\psi}_{n\mathbf{k}}(\mathbf{r})$, that match the exact ones outside a certain radius around each nucleus:

$$\psi_{n\mathbf{k}}(\mathbf{r}) = \tilde{\psi}_{n\mathbf{k}}(\mathbf{r}) - \sum_{R} \left[\tilde{\psi}_{n\mathbf{k}}^{1,R}(\mathbf{r}) - \psi_{n\mathbf{k}}^{1,R}(\mathbf{r}) \right]. \tag{4}$$

In this method the one-center pseudo-wavefunctions $\tilde{\psi}_{n\mathbf{k}}^{1,R}(\mathbf{r})$ and the exact one-center wavefunctions $\psi_{n\mathbf{k}}^{1,R}(\mathbf{r})$ are represented on radial grids. A schematic picture of the procedure is given in Fig. 1.

Regarded as a generalized eigenvalue problem the Kohn-Sham equations and can be solved very efficiently using iterative methods. The diagonalization can be efficiently parallelized, since equation (2) is diagonal in the index n of the eigenstate ("inter-band-distribution"); furthermore, if there are enough nodes available, the diagonalization for the n-th state may be parallelized as well ("intra-band-distribution"). However, a limiting factor is the communication overhead required for the redistribution of the wavefunctions between all nodes, which is necessary during the orthogonalization procedure of the eigenstates.

We use the DFT-PAW implementation in the Vienna Ab-initio Simulation Package (VASP) [5], together with the gradient-corrected parametrization of the exchange-correlation energy [6,7]. The Kohn-Sham matrix is diagonalized using the Residual Minimization Method with Direct Inversion in Iterative Subspace (RMM-DIIS) [8]. This scheme is preferred over the more common Conjugate Gradient (CG) method, since the latter requires explicit orthonormalization of the search vector for each wavefunction with respect to all other wavefunctions during each iteration step, an $\mathcal{O}(N^3)$ operation. The RMM-DIIS scheme reduces the number of $\mathcal{O}(N^3)$ operations to a minimum [5]. Parallelization is done using the Message Passing Interface (MPI).

2.2 Modeling of Non-periodic Structures

The expansion of wavefunctions into a set of plane waves is very efficient for periodic structures like infinite crystals. However, many systems of special interest are partially (surfaces, nanowires) or completely non-periodic (nanodots). To model such systems we are using the periodically repeated supercell approach.

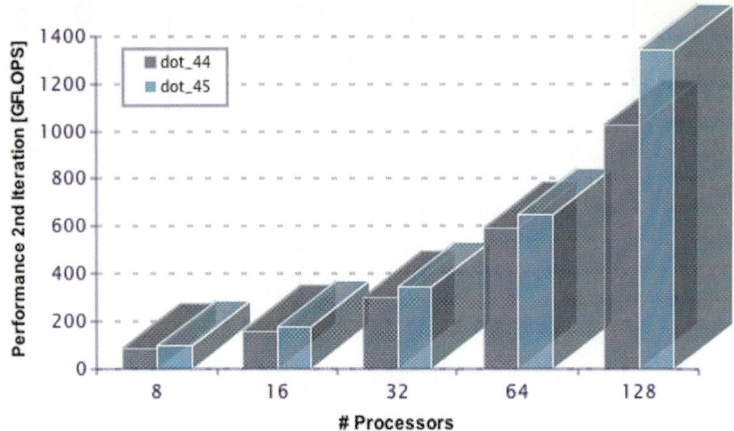

Fig. 2. Performance on the NEC SX-8 for two different PbTe nanodot systems: dot_45 and dot_44, containing 1000 and 512 atoms, respectively [10]

In the case of PbTe/CdTe interfaces each super-cell consists of two material slabs. Both of them containing 14 atomic layers (or 7 neutral bilayers) of PbTe or CdTe. Within the interface plane we use (2×1) interface supercells. Brillouin zone (BZ) integrations are performed on regular meshes in reciprocal space [9]. Wavefunctions are expanded into plane waves up to a cutoff energy of 200 eV. Relaxations of ionic positions are carried out using conjugate gradient or quasi-Newton algorithms, until the Hellmann-Feynman forces fall below 20 meV/.

For embedded or free-standing PbTe nanodots we are using super-cells from $(25.64\text{Å}\times25.64\text{Å}\times25.64\text{Å})$ up to $(38.46\text{Å}\times38.46\text{Å}\times38.46\text{Å})$. Due to the vanishing dispersion in k space just the Γ-point is used for the BZ sampling, which speeds up the calculations considerable.

2.3 Computational Cost

A large part of our calculations were carried out on the NEC SX-8 system and a Cray Xb1 Opteron cluster. Figure 2 shows the scaling behaviour of our code on the NEC Sx8 system for an embedded PbTe nanodot. The testruns presented here do only the first ionic step while a production run typically performs some 100 ionic steps. The performance for the iteration part is computed as the difference between a complete run for two ionic steps one ionic step. The computation is dominated by complex matrix-matrix multiplication (CGEMM). The sustained iteration performance for both cases exceeds 1 TFLOPS already on 16 nodes NEC SX-8 (Fig. 2). The sustained efficiency is between 79 and 50 % [10].

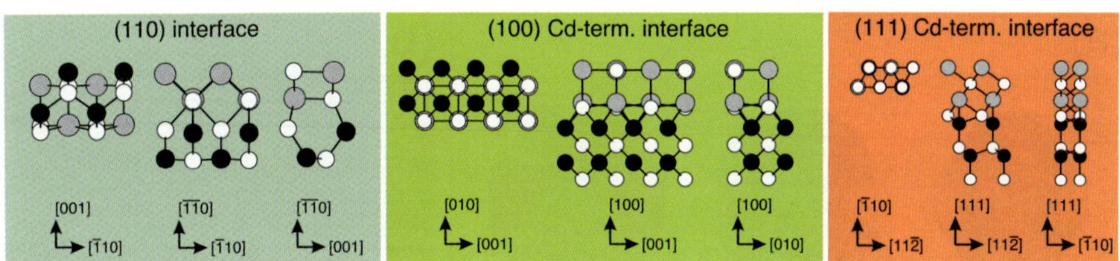

Fig. 3. Schematic representation of the atomic arrangements at the relaxed PbTe/CdTe interfaces (left panel: (110) face, middle panel: (100) face, and right panel: (111) face). In the case of the polar (100) and (111) faces only the Cd-terminated interfaces. The atoms are represented by open (Te), filled (Cd), and shaded (Pb) circles

3 Results

3.1 Surface Structure of PbTe/CdTe Interfaces

Figure 3 shows schematic ball-and-stick representations of the relaxed geometries of the (100), (110) and (111) PbTe/CdTe interfaces. The combination of the polar zincblende(zb)-CdTe(100) and (111) surfaces with the corresponding rocksalt(rs)-PbTe surfaces gives rise to interfaces with different terminations of the zb-crystal. These interfaces are completely different with respect to extent (averaged distance of the outermost atomic planes of the two crystals) and bonding behavior. At (100)/(111) interfaces with a Cd-termination of the zb-crystal we find an interface extent of nominal $(a_0/4)/(\sqrt{3}\,a_0/12)$, whereas the extent at Te-terminated interfaces is found to be $(a_0/2)/(\sqrt{3}\,a_0/4)$, being in good agreement with recent HRTEM investigations [1]. In both cases only displacements parallel to the interface normal occur. The main effect is given by some kind of rumpling, well known from the free PbTe(100) surface [11]. At the non-polar (110) interface the rumpling effect results in a weak splitting of the electrostatic neutral and stochiometric PbTe(110) planes into bilayers. The effect vanishes in the slab center and increases toward the interfaces. The split bilayers change the polarity in an oscillating way. But the most prominent effect at this kind of interfaces is an offset of 0.38 Å in the [001] direction (parallel to the interface) between the PbTe and CdTe crystals. This can be interpreted as the result of two tendencies. First of all the Cd atoms in the first interface layer (CdTe side) try to occupy a fourfold-coordinated site and secondly there are repulsive forces between the Pb and Te ions in the first interface layers at PbTe and CdTe side, respectively. Using the obtained structural properties, the averaged interface energies per area A, $\gamma(hkl) = E_{\text{free}}^{\text{inter}}(hkl)/A$ can be calculated. The dipole corrected values for the (110), (100), and (111) PbTe/CdTe interfaces are 0.20, 0.23, and 0.19 J/m^2. Hence, the (110) and (100) interfaces, respectively, are the least stable from the energetical point of view. However, all three interface energies are roughly close to their average value 0.2 J/m^2. But in contrast to the corresponding free surfaces there is

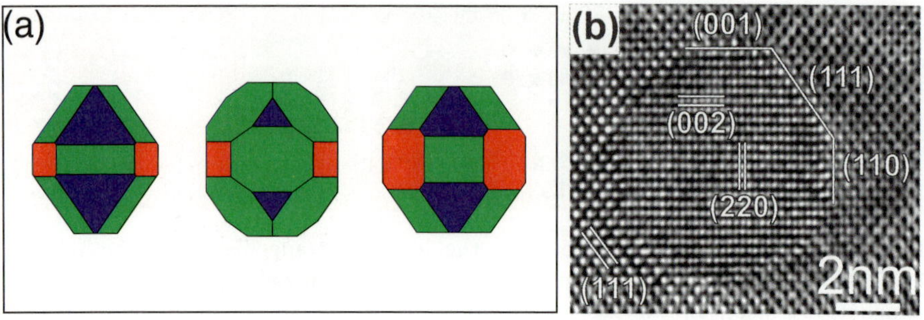

Fig. 4. (a) Projection of ECSs along the [110] zone axis, with the abscissa along [1$\bar{1}$0] and the ordinate along [001]. The green facets represent {110}, the red facets {100} and the blue facets {111} faces. The ECS at the left is constructed using the calculated values, for the ECS at the middle we have changed the (111) interface energy to 0.22 J/m^2, and the ECS at the right is constructed using equal interface energies of 0.2 J/m^2. (b) HRXTEM image of a PbTe quantum dot in CdTe matrix, projected onto same crystal plane [1]

no clear physical trend. In agreement with the surface polarity and the number of dangling bonds the surface energies show a relatively clear ordering $\gamma(110) < \gamma(111) < \gamma(100)$ for zincblende (zb) and $\gamma(100) < \gamma(110) < \gamma(111)$ for rocksalt (rs). The same order of surface energies versus orientations were predicted by other authors for zb-InAs and rs-NaCl. [12,13] The difference in energy ordering of the two sequences does not suggest a unique picture for rs and zb interfaces, but it confirms the weak variation with the orientation of the average interface energies.

The equilibrium crystal shape (ECS) of embedded PbTe nanocrystallites can be obtained by means of a Wulff construction using the interface energies. The results are presented in Fig. 4 (a). They demonstrate a remarkable sensitivity of the shape with respect to tiny variations of the interface energies. Concerning the three interface orientations [110], [100], and [111], we always find a rhombo-cubo-octahedral ECS. Only the relative ratios of the areas of the {110}, {100}, and {110} facets vary with the absolute values of the interface energies. In any case, Fig. 4 (a) clearly indicates that the three interfaces {110}, {100}, and {111} are thermodynamically stable.

The (110) projections of the ECSs shown in Fig. 4 (a) are compared to a HRTEM image of a real PbTe quantum dot embedded in CdTe matrix [Fig. 4 (b)] [1]. The projections indicate the best agreement for a Wulff construction with interface energies of 0.20 J/m^2, 0.23 J/m^2, and 0.22 J/m^2 for the {110}, {100}, and {111} facets, respectively. Considering the strong sensitivity of the ECSs on the calculated interface energies this is in good agreement with our theoretical predictions. However, the energy contribution due to the edges and vertices can have a considerable influence on the stability of small nanocrystallites. Hence, for more accurate predictions of the ECS of small PbTe quantum dots, one has to perform explicit *ab initio* calculations

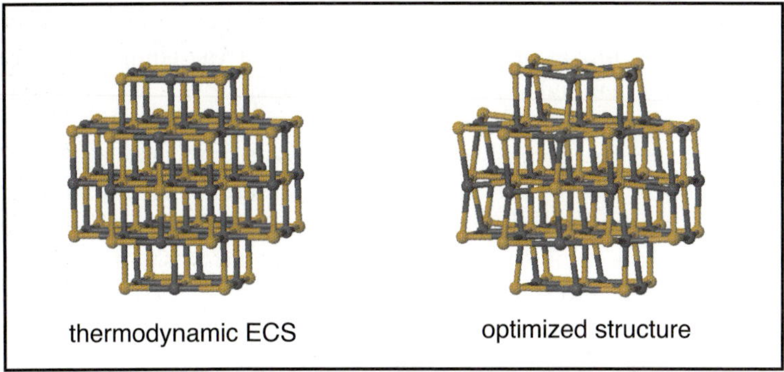

Fig. 5. PbTe nanodot with a diameter of 12.82 embedded in CdTe matrix (not shown). Left hand side thermodynamic ECS obtained via Wulff construction, right hand side relaxed structure

for PbTe quantum dots embedded in a CdTe matrix, or at least, to model the system with capped nanocrystallites as suggested in Ref. [14].

3.2 Embedded PbTe Nanodots

As starting structure for an *ab initio* structural relaxation of the embedded PbTe-dots, we use the ECS obtained by a Wulff construction. To minimize eventually artificial interactions via the periodic boundary conditions, we have embedded a PbTe nanodot with a diameter of 12.82 Å in a CdTe matrix of differnt sizes. The obtained results for the atomic positions are nearly independent of the matrix size, thus confirming the accuracy of our approach. In Fig. 5 a schematic stick and ball representation of the relaxed structure is given. These calculations confirm the stability of the thermodynamic predicted ECS. At the dot-matrix interfaces a strong rumpling effect can be observed. This is in good agreement with findings for the free standing PbTe/CdTe interfaces (of corresponding orientations) as discussed in the last section.

Using these atomic configurations we are able to calculate electronic properties like e.g. the spacial and energetic localization of the highest occupied and lowest unoccupied states. These informations can than be used to investigate optical properties, in particular the photoluminescence spectra, because those are of special interest for future applications like e.g. a mid infrared quantum-dot laser.

4 Summary

In conclusion, we calculated the structural properties of (110), (100) and (111) PbTe/CdTe interfaces, which consist of two different cubic crystal structures with partially ionic bonds. A thermodynamic method as well as a *first principles* structural optimization was used to construct the equilibrium shape of a PbTe nanocrystal embedded in a CdTe matrix.

Our results predict for the ECS a rhombo-cubo-octahedron with 26 interface facets. The relative size of the facet areas for different orientations depends sensitively on the actual relative values of the interface energies. The optimized dot-matrix interfaces exhibit nearly the same behavior as the investigated free standing PbTe/CdTe interfaces.

Since the nano-objects, dots and surfaces, are modelled using repeated supercells, many atoms and electrons are involved. We have cells up to 1.000 atoms. A reasonable treatment is therefore only possible with a well parallized code. The corresponding performance of VASP has been demonstrated for the NEC SX-8 system.

The work was financially supported through the Fonds zur Förderung der Wissenschaftlichen Forschung (Austria) in the framework of SFB25, Nanostrukturen für Infrarot-Photonik (IR-ON) and the EU NANOQUANTA network of excellence (NMP4-CT-2004-500198). Grants of computer time from the Höchstleistungsrechenzentrum Stuttgart are gratefully acknowledged.

References

1. W. Heiss *et al.*, Appl. Phys. Lett. 88, 192109 (2006)
2. P. Hohenberg and W. Kohn, Phys. Rev. **136**, B864 (1964).
3. W. Kohn and L. Sham, Phys. Rev. **140**, A1133 (1965).
4. P. E. Blöchl, Phys. Rev. B **50**, 17953 (1994).
5. G. Kresse and J. Furthmüller, Phys. Rev. B **54**, 11169 (1996).
6. J. P. Perdew and A. Zunger, Phys. Rev. B **23**, 5048 (1981).
7. J. P. Perdew, J. A. Chevary, S. H. Vosko, K. A. Jackson, M. R. Pederson, D. J. Singh, and C. Fiolhais, Phys. Rev. B **46**, 6671 (1992).
8. P. Pulay, Chem. Phys. Lett. **73**, 393 (1980).
9. H. J. Monkhorst and J. D. Pack, Phys. Rev. B **13**, 5188 (1976).
10. S. Haberhauer, NEC - High Performance Computing Europe GmbH, Nobelstrasse 19, D-70569 Stuttgart, Germany, **SHaberhauer@hpce.nec.com**
11. A. A. Lazarides, C. B. Duke, A. Paton, and A. Kahn, Phys. Rev. B **52**, 15264 (1999).
12. N. Moll, M. Scheffler, and E. Pehlke, Phys. Rev. B **58**, 4566 (1998).
13. D. Wolf, Phys. Rev. Lett. **68**, 3315 (1992).
14. L. E. Ramos, J. Furthmüller, and F. Bechstedt, Phys. Rev. B **72**, 045351 (2005).

Environment/Climate Modeling

The Agulhas System as a Key Region of the Global Oceanic Circulation

Arne Biastoch[1], Claus W. Böning[1] and Fredrik Svensson[2]

[1] Leibniz-Institut für Meereswissenschaften,
Düsternbrooker Weg 20, 24106 Kiel, Germany
`abiastoch|cboening@ifm-geomar.de`
[2] NEC High Performance Computing Europe GmbH,
Heßbrühlstraße 21B, 70565 Stuttgart, Germany
`fsvensson@hpce.nec.com`

1 Abstract

The Agulhas system at the interface between the Indian and Atlantic Ocean is an important region in the global oceanic circulation with a recognized key role in global climate and climate change. It is dominated by high temporal and horizontal variability. This project aims to realistically simulate this complex current system and its effect on the interoceanic transport with the highest spatial resolution to date. Using a hierarchy of global ocean models with realistic and idealized atmospheric forcing, the effect of inter-ocean transport on the large-scale circulation in the Atlantic will be established. This includes the variability of the meridional overturning and heat transports. The core of the project is a high-resolution model of the Agulhas region that is nested in a global one at lower resolution. Both models are able to interact, which allows one to study the feedbacks from the high-resolution model on the large-scale circulation. This project is embedded in the European ocean modeling effort *DRAKKAR*.

2 Scientific Objective

The meridional overturning circulation, consisting of the southward transport of cold deep water that has been formed in the subpolar North Atlantic and the compensating warm surface flow, determines large-scale transports of heat, freshwater and dissolved anthropogenic trace gases, and plays an important role in the global climate system. A quantitative understanding of the factors determining the physics and strength of this circulation and its variability on different time scales is therefore one of the important fields of oceanographic research and a prerequisite for the realistic simulations in climate models.

The mechanisms and influences of fluctuations of the deep water formation in northern latitudes in response to interannual to decadal variability of the surface forcing has been addressed in several model and observation studies (e.g. [6]). In contrast there are only few systematic studies on the variability of the return transport of surface waters.

For the return transport competing theories exist regarding the inflow into the South Atlantic: the cold water route which involves to the transport of intermediate water through the Drake Passage, and the warm water route, fed by the interoceanic exchange of surface water between Indian Ocean and Atlantic, through the Agulhas system. Although the relative portion of both paths in the time-mean is still under discussion in analyses from observations, models and inverse calculations (an actual summary is in [4]) , it is accepted that the signal from the Agulhas system is strongly fluctuating and therefore extremely difficult to quantify. The variability of the upper branch of the thermohaline circulation is therefore mostly attributed to this path.

The circulation in the Agulhas system is strongly determined by different time and space scales: The Agulhas Current [7], flowing southward close along the African continent as one of the most intense western boundary currents, is very stable and has only minor seasonality. In contrast its retroflection back into the Indian Ocean is, as a result of its geographical position subject to strong seasonal and interannual fluctuations. Also the transport of warm and salty water into the South Atlantic is mainly taking place in mesoscale eddies (Agulhas rings), determined by strong time and space variability [2] and therefore difficult to observe. [5] has already pointed out that the inflow of warm and salty water should act on the strength of the meridional overturning due to the dynamical effect of the preconditioning of deep water formation in the subpolar North Atlantic. At the same time the interoceanic transport is strongly dependent on local conditions, such as the wind characteristics in the Indian Ocean, and reacts sensitively to small changes [2]. Paleo-oceanographic analyses for example point out that in glacial times the transport from the Indian Ocean to the Atlantic was reduced to a minimum (e.g. [10]).

For climate studies it has already been stated in the IPCC Third Assessment Report, that the transport by Agulhas rings could generate variations in the thermohaline circulation [11]. Due to the lack of horizontal resolution (and therefore the exclusion of explicitly simulated Agulhas rings and small-scale processes) in the existing climate models this has not been proven so far. It has been shown only in very simple or coarse resolving models that the transport of heat and salt through the Agulhas region has a substantial effect on the strength of the meridional overturning circulation in the Atlantic as a whole [12]. But it remains unclear how significant such findings are due to the use of very coarse resolution (and therefore the neglect of mesoscale features).

3 Model Hierarchy

The project consists of a suite of global and regional ocean models based on the recent version 9 of the OPA model [8], now called *NEMO*. NEMO is a state-of-the-art three-dimensional ocean model coupled to a dynamic-thermodynamic sea-ice model. The ocean component is based on so-called "primitive equations", stepping velocity, temperature and salinity forward in time. A free surface formulation (e.g. by a conjugate gradient solver), a polynominal fit of the density equation and lots of parameterization of different ocean physics sum up to modular unit of different numerical methods. It is written in FORTRAN90, has a finite differences layout with a horizontal (geographical) domain decomposition for MPI parallelization.

The first global configuration has a nominal horizontal resolution of $1/2°$ (*ORCA05*), is therefore coarse-resolving and does not contain the detailed physics of the Agulhas system. Its use is to first-hand test the response of the large-scale circulation to atmospheric conditions (Fig. 1) and to motivate higher-resolved, more costly studies. It also represents the target resolution of future climate models and will therefore give an assessment of the reproduction of the Agulhas system in this class of models.

The second configuration is with $1/4°$ (*ORCA025*) already eddy-permitting [1] and does contain the larger Agulhas rings and eddy structures (Fig. 2). It will also introduce the effect of such mesoscale structures on the large-scale

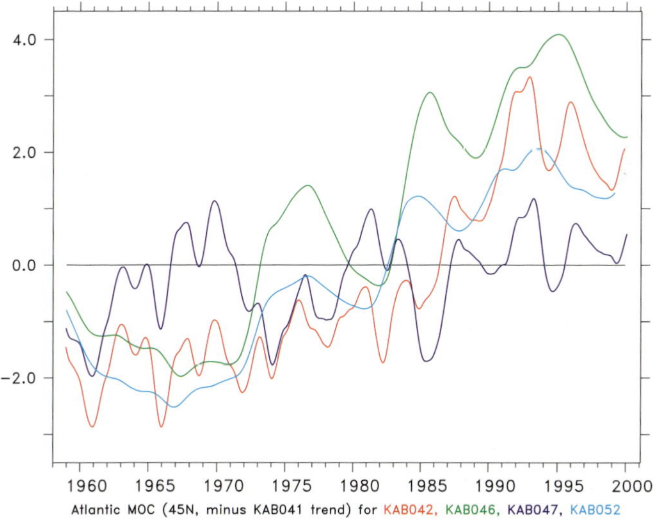

Fig. 1. Strength of the North Atlantic meridional overturning circulation at 45° N, a measurement of the strength of the global thermohaline circulation in ORCA05. Shown are different experiments with full forcing variability (red), compared to heat and freshwater variability only (green), wind variability only (blue) and heat variability only (lightblue)

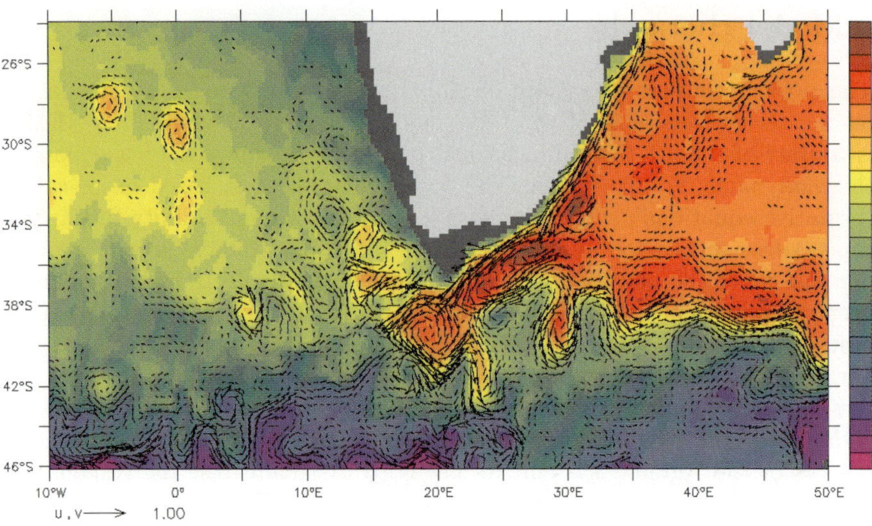

Fig. 2. Circulation around South Africa in ORCA025. Shown is a snapshot of temperature and velocity in 450 m depth. Close to the African coast the Agulhas Current is transporting warm water southward. After overshooting the southern tip of the continent it retroflects back into the Indian Ocean, pinching off Agulhas rings that are transporting the warm and salty water northward in the South Atlantic

circulation. The two global configurations will both be integrated at HLRS (SX-8) and DKRZ Hamburg (SX-6).

The full representation of the Agulhas system can only be reached by simulating the region at high spatial resolution. The old dilemma of systematic model studies, either by simulating the important processes through adequate resolution – by limiting to regional models and/or restricting to short time scales – or by simulating the dynamics through parameterization of the small-scale phenomena, has been lead to the use of global, eddy-resolving models with $1/10°$ resolution (e.g. [9]); The vast computational cost prevents dedicated experiments to investigate the response of the system. Here we are taking another approach to examine the effect of the Agulhas model more selectively: by two-way nesting a regional high-resolution model into a more efficient global model and still to determine the relative importance of the Agulhas Current system and its mesoscale variability. For this purpose it is planned to nest $1/10$ - $1/12°$ regional models into the two global models (Fig. 3). The nested models are able to interact with their hosts and allow one to study feedbacks from the high-resolution model on the large-scale circulation. The nesting approach *Adaptive Grid Refinement In Fortran, AGRIF*, [3] is realized via a preprocessing step, inserting pointer directives into the model code. These pointers allow both models to interact in a two-way coupling where model variables from the host update the boundaries of the nest and model variables from the nest update the coarser resolved grid points of the

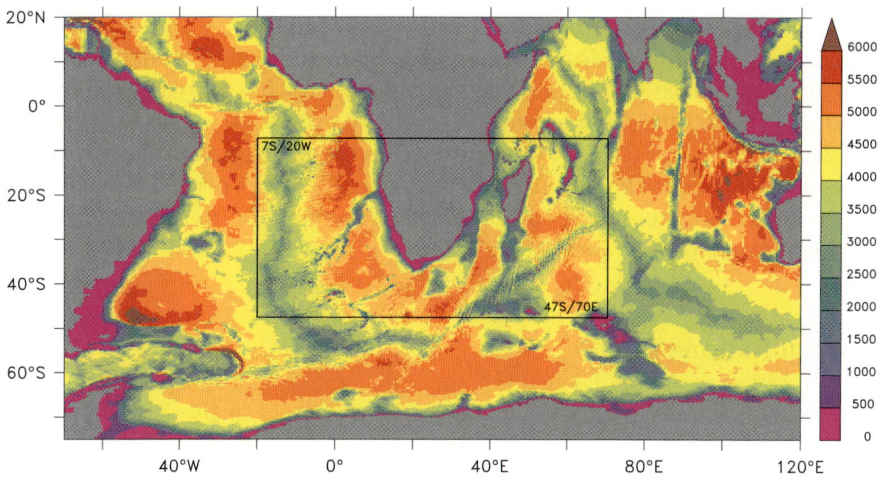

Fig. 3. Part of ORCA05 bathymetry (color shows water depth in meters) with boundaries and bathymetry of an 1/10° Agulhas nest

host. This configuration is very large in computational cost and requires a multi-node supercomputer, it will run on the HLRS SX-8 only.

4 Computing Requirements

With $722 \times 511 \times 46 \approx 17 \times 10^6$ grid points (lon $\times$ lat $\times$ depth) and a time step of 40 minutes a typical experiment length of 45 years ORCA05 takes 1,300 CPU hours (unoptimized, one node SX-8 with autotasking). Compared to that ORCA025 has a doubled resolution (factor 4 in grid points) and – due to higher resolved physics – a halved time step (20 min). A typical experiment will therefore end up with more than 10,000 CPU hours. A nested ORCA025-AGRIF model will then easily reach 30,000 - 40,000 CPU hours per experiment.

Given the early phase of this project within the TERAFLOP first vector and I/O performance improvements have been made. The current performance is 4 GFLOP/s per processor (on $3 \times 8 = 24$ processors). With 25% of the peak performance this is a reasonable starting figure for a GFD code. Further optimization is currently in progress, concentrating on the MPI parallelization and the most consuming routines, the ice calculation and the external solver. "Most consuming" means 7 - 10% of the total, underlining the heterogeneous structure of the code with no clear top-level routine.

Beside the "pure number crunching" figure, the massive output is a typical bottleneck for ocean and climate models: Already ORCA05 has a total output of 405 GB per experiment which is significant taking into account the relatively coarse resolution. To explore the time evolution of the higher-resolved oceanic fields (velocity, temperature and salinity) in ORCA025 does increase significantly towards 6 TB per experiment. Since the nested model does contain almost as much grid points as its simultaneously integrated global model (ORCA05, ORCA025) and has to be performed even with a shorter time step another factor 2-3 in terms of computing time can be expected. Similar to the other configurations the output will easily exceed 10 TB per experiment and will therefore require special optimizations of the input/output technique and the long-term storage.

5 Conclusions

The described model framework to study the dynamics in the Agulhas region and its feedbacks on the large-scale circulation is based on a nesting approach to avoid the typical dilemma of high spatial resolution versus long-term integrations. In terms of performance the model code is at a good starting point, already using a quarter of the peak performance in its current version. A special challenge for hardware and storage will be the vast amount of data generated and its storage on a long-term basis.

References

1. Barnier, B., Madec, G., Penduff, T., Molines, J.-M., Treguier, A.-M., Beckmann, A., Biastoch, A., Böning, C., Dengg, J., Gulev, S. Le Sommer, J., Remy, E., Talandier, C., Theetten, S., and Maltrud, M. (2006). Impact of partial steps and momentum advection schemes in a global ocean circulation model at eddy permitting resolution. *Ocean Dynmics*, page in press.
2. De Ruijter, W. P. M., Van Leeuwen, P. J., and Lutjeharms, J. R. E. (1999). Generation and evolution of Natal Pulses, solitary meanders in the Agulhas Current. *J. Phys. Oceanogr.*, 29:3043 – 3055.
3. Debreu, L. (2000). *Raffinement adaptif de maillage et methodes de zoom. Applications aux modeles d'ocean*. PhD thesis, University Joseph Fourier, Grenoble.
4. Donners, J. and Drijjhout, S. S. (2004). The Lagrangian view of South Atlantic interocean exchange in a global ocean model compared with inverse model results. *J. Phys. Oceanogr.*, 34:1019 – 1035.
5. Gordon, A. L., Weiss, R. F., Smethie, W. M., and Warner, M. J. (1992). Thermocline and intermediate water communication between the South Atlantic and Indian Oceans. *J. Geophys. Res.*, 97:7223 – 7240.
6. Hurrell, J., Kushnir, Y., Ottersen, G., and Visbeck, V., editors (2003). *The North Atlantic Oscillation: Climate Significance and Environmental Impact*, volume 134 of *Geophysical Monograph Series*. AGU.
7. Lutjeharms, J. R. E. (1996). The exchange of water between the South Indian and South Atlantic oceans; a review. In Wefer, G., editor, *The South Atlantic Present and Past Circulation*. Elsevier.

8. Madec, G., Delecluse, P., Imbard, M., and Levy, C. (1999). Opa 8.1 ocean general circulation model reference manual. Technical report, Institut Pierre Simon Laplace des Sciences de l'Environment Global.

9. Matsumoto, K., Sarmiento, J., Key, R., Aumont, O., Bullister, J., Caldeira, K., Campin, J.-M., Doney, S., Drange, H., Dutay, J.-C., Follows, M., Gao, Y., Gnanadesikan, A., Gruber, N., Ishida, A., Joos, F., Lindsay, K., Maier-Reimer, E., Marshall, J., Matear, R., Monfray, P., Mouchet, A., Najjar, R., Plattner, G.-K., Schlitzer, R., Slater, R., Swathi, P., Totterdell, I., Weirig, M.-F., Yamanaka, Y., Yool, A., and Orr, J. (2004). Evaluation of ocean carbon cycle models with data-based metrics. *Geophysical Research Letters*, 31(7):L07303 1–4.

10. Peeters, F. J. C., Acheson, R., Brummer, G.-J. A., De Ruijter, W. P. M., Ganssen, G. G., Schneider, R. R., Ufkes, E., and Kroon, D. (2004). Vigorous exchange between Indian and Atlantic Ocean at the end of the last five glacial periods. *Nature*, 400:661–665.

11. Stocker, T. F., Clarke, G. K. C., Le Treut, H., Lindzen, R. S., Meleshko, V. P., Mugura, R. K., Palmer, T. N., Pierrehumbert, R. T., Sellers, P. J., Trenberth, K. E., and Willebrand, J. (2001). Physical climate processes and feedbacks. In Houghton, J. T., Ding, Y., Griggs, D. J., Noguer, M., van der Linden, P. J., Dai, X., Maskell, K., and Johnson, C. A., editors, *Climate Change 2001: The Scientific Basis. Contribution of Working Group I to the Third Assessment Report of the Intergovernmental Panel on Climate Change.* Cambridge University Press.

12. Weijer, W., De Ruijter, W. P. M., and Dijkstra, H. A. (2001). Stability of the Atlantic overturning circulation: Competition between Bering strait freshwater flux and Agulhas heat and salt sources. *J. Phys. Oceanogr.*, 31:2385–2402.

ECHAM5 – An Atmospheric Climate Model and the Extension to a Coupled Model

Luis Kornblueh

Max-Planck-Institute for Meteorology
Bundesstr. 53,
D-20146 Hamburg, Germany.
luis.kornblueh@zmaw.de

1 Introduction

Science around the Earth's climate is dealing with physical, chemical, and biological processes, as well as human behavior contributing to the dynamics of the Earth system, and specifically how they relate to global and regional climate changes. The major target is to observe, monitor, analyze, understand, and predict the Earth system in order to improve it's management.

Since the middle of the 19th century the temperature of the Earth's surface has increased by almost 1 K. Much of this observed global warming is due to human activities. Recent climate simulations suggest that the global, annual mean temperature increases by 2.5 K to 4 K at the end of the 21st century, if emissions of carbon dioxide and other greenhouse gases continue to grow unabatedly. The most important results of contemporary studies are summarized as follows:

- Land areas will warm more rapidly than the oceans. The most notable warming is expected at high northern latitudes, in particular in the Arctic region.
- The precipitation amount tends to increase in humid climate zones (tropics, middle and high latitudes) and decreases in arid climate zones (subtropics).
- The precipitation intensity and risk of flooding increase in most regions.
- In most parts of Europe the snow amount in winter decreases by 80-90% until the end of this century. A decrease by 30-40% is simulated for the Alps and for the Norwegian mountains.
- The length of dry spells increases world-wide. The risk of drying is most pronounced in the Mediterranean countries, in South Africa, and in Australia.

- At the end of this century the contrast between dry and wet climate zones becomes more pronounced, and precipitation extremes of both signs are increasing.
- The intensity of winter storms increases in Central Europa but decreases in the Mediterranean area.
- Due to thermal expansion, the global sea-level rises by 20 cm to 30 cm until the end of this century. The melting of Greenland ice contributes to some 15 cm, whilst enhanced snowfall in Antarctica tends to decrease the global sea-level by 5 cm.
- The pronounced warming of the Arctic leads to thinner sea-ice in winter and smaller sea-ice extent in summer. The observed loss of summer ice in recent years is expected to continue in the climate projections: Until the end of this century the whole Arctic Ocean will become ice-free in summer.
- Higher temperatures and precipitation amounts reduce the density of the surface water in the North Atlantic and, hence, the strength of the thermohaline circulation and the northward heat transport. However, the weakening of the circulation by about 30% until the end of this century has little effect on t he European climate, which continues to warm due to higher levels of atmospheric greenhouse gases. In the past, the greenhouse warming has partially been masked by increasing atmospheric concentrations of anthropogenic aerosols like sulfate and black carbon. Drastic measures to improve air quality would result in a rapid global warming of almost 1 K within ten years. Thus, strategies to limit climate warming below a specified threshold need to be reconciled with strategies to reduce air pollution.

During the last 150 years we have introduced massive perturbations in some of the climate's main forcing variables (Fig. 1). How will the system respond to such forcings?

2 Modelling the Climate System

We have to deal with a very complex system as shown in Fig. 2. It is a highly non-linear dynamic system. In nonlinear complex systems, minute actions can cause long-term, large-scale changes. These changes can be abrupt, devastating, surprising, and unmanageable (see Fig. 3).

The key questions to be raised determine the model to be build. The key questions we currently try to answer can be summarized as: Are there critical thresholds that cause abrupt climate change? What are the processes that regulate the variability of CO2 concentration? What controls atmospheric ozone and oxidation processes? Which processes control atmospheric aerosols and aerosol interaction with clouds and climate? Which are the most vulnerable regions and sectors under global change and why? Is the Earth system manageable at all in terms of long-term *climate steering*?

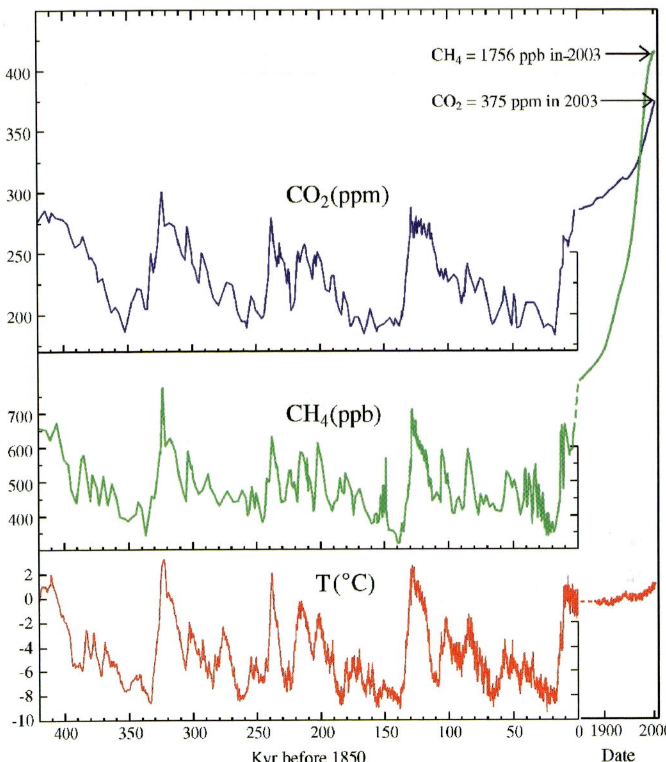

Fig. 1. Concentrations of some greenhouse gases over the last 400 thousand years

The system to look for is a full Earth System Model (ESM), which does contain all relevant compartments as shown in Fig. 4. General strategy is to split the whole system into components which can be handled by different scientific communities.

3 The Atmospheric Climate Model ECHAM5

The climate model ECHAM5 consists of the atmospheric component of an ESM. It's basic version consists of the following components:

- Spectral dynamical core
- Semi-implicit leapfrog time differencing
- Flux-form semi-Lagrangian transport of passive tracers
- Shortwave radiation based on δ-Eddington approximation
- Longwave radiation based on the correlated-k method

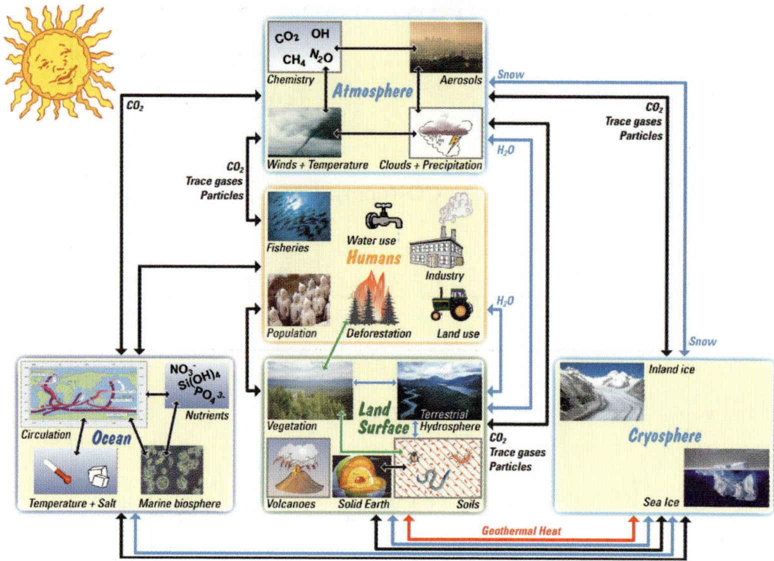

Fig. 2. The Earth's complex climate system

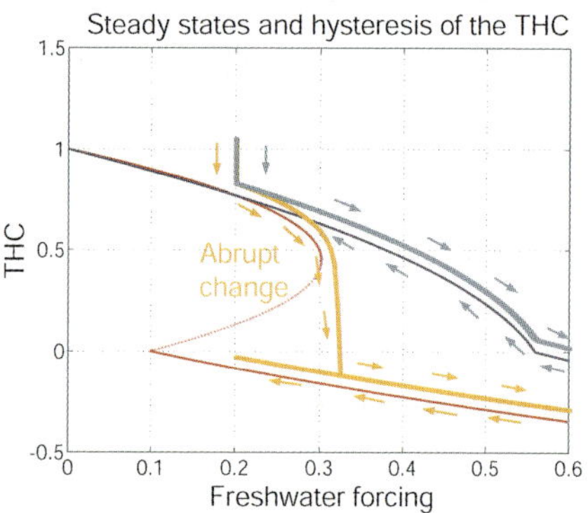

Fig. 3. State space development with potential abrupt changes

- Stratiform clouds based on micro-physical prognostic equations for water vapour, water, and ice
- Convection solved by a mass-flux scheme based on Tiedtke
- Subgrid-scale induced gravity wave drag
- Vertical diffusion (subgrid-scale turbulence closure by TKE)
- Solving the surface energy balance equation

The process interactions in the model are shown in Fig. 5.

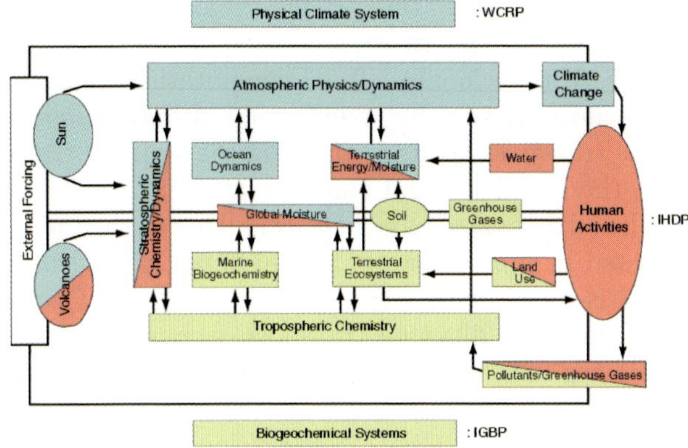

Fig. 4. Compartments of an Earth System Model

4 Internal Model Organization, Parallelization and Optimization Strategies

4.1 MPI Parallelization

ECHAM5 is organized in several model states which allow computations to be performed in an optimal way with respect to the properties of each space.

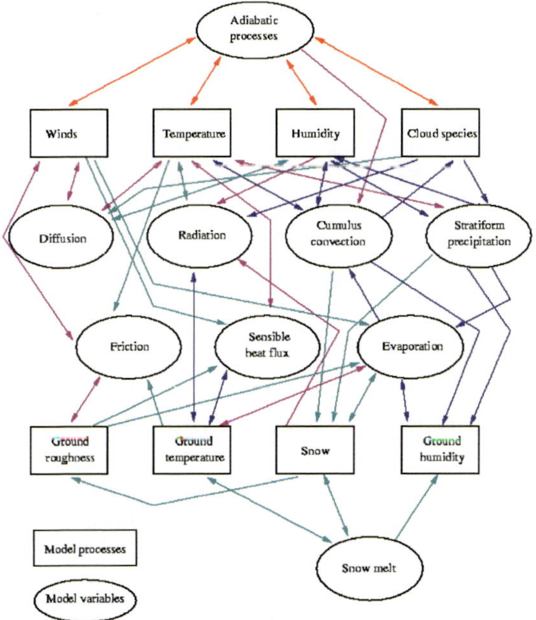

Fig. 5. Interaction between different parts of the atmospheric compartment of an ESM

Solutions based on spherical harmonics describing large parts of the general PDE's to be solved on the sphere are organized in spectral space for handling the respective spherical harmonics. To do computations representing the process interactions those have to be transformed into grid point space. This is done by a double transpose strategy splitting the spectral transform into the required parts – the Legendre- and Fourier-transform. An intermediate space is required for the advection of passive tracers. Due to the fact that the whole model is integrated in time based on a semi-implicit integration scheme the timestep allowed is introducing wind speed exceeding a Courant-number of one in east-west direction allowing transport over much more than a single grid-cell per timestep. The required halo for the parallelization is hardly predictable so that the model state is restructured in a way that all latitude bands are located on a single CPU. In north-south direction the wind-speed is small enough to allow a static halo of one. To achieve sufficient granularity this state is splitted over levels and tracers instead of longitudes. This reorganized for the process interaction computations, which do require vertical columns. After finalizing the grid-point computations the transform back into spectral spaces takes place following a double-transpose strategy again.

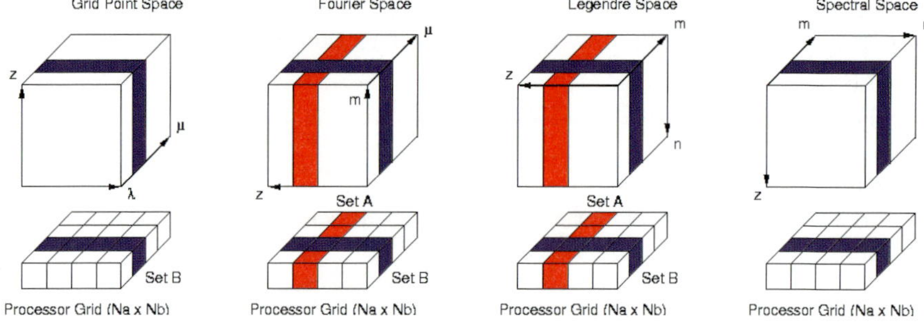

Fig. 6. The models decomposition in 2 dimensions and transpositions

5 Loop-level Parallelization – OpenMP

To allow for more improved usage of local SMP based memory below the MPI parallelization layer an additional parallelization based on OpenMP is used. The code is structured in a way that most of the OpenMP directives can be applied on a very high level without interfering with code handled by scientists developing the model. The process interaction computations are handled by OpenMP orphaning. The remaining code is handled by OpenMP on loop level. This layer of parallelization has been developed primarily with vector-architecture in mind. The development platform is a NEC SX-6 system. Two major problems have been encountered during the development. One is the implementation of OMP WORKSHARE which does not perform at all. A second problem is so called OS jitter (see Sect. 7.3).

6 Vectorization and Adaptation for Cache-based Microprocessor Systems

This part of optimization is the most difficult one. If portability across architectures and maintainability is important for an application as it is the case for ECHAM5, compromises have to made.

The process interaction computational block is optimized by high level strip-mining. Using a namelist definable variable the innermost loop length can be set. A natural choice for vector machines is a multiple of the vector-register length. ON cache based machines one can measure the runtime and determine an optimal value, which is only changing for different vertical model resolutions. Improvements can be gained in the order of 2 and more. The calculations in spectral space can be solved by *fast and dirty* code by intentional collapsing dimensions. Clean code is provided for each of this optimizations to allow for bounds-checking.

As well the Legendre as the Fourier-transformation are highly optimized. The Legendre transformations are implemented in terms of DGEMM calls. The Fourier-transform used is faster than almost all other available in high-performance libraries and platform independent.

A major performance bottleneck is the tracer transport because of suffering from sufficient vector-length on vector-architectures, while running quite acceptable on other platforms.

7 Some Remarks on the Performance on the NEC SX-6

7.1 Intranode Performance Degradation

By using a perfectly parallel code the degradation by technical issues of a single SMP node can be determined. Therefore ECHAM5 is spawned as single-CPU instance by MPI on a single node. The runtime on a single CPU on an exclusive node has been accounted an efficiency of one. Successively increasing the number of CPUs (single CPU models running concurrently without any communication) is used to determine the influence of a single node system on the performance of a MPI application. The performance loss is already close to 15% (Fig. 7), which can be purely denoted to the systems loss. If there is any random component in this loss, scaling over several nodes will be strongly degraded.

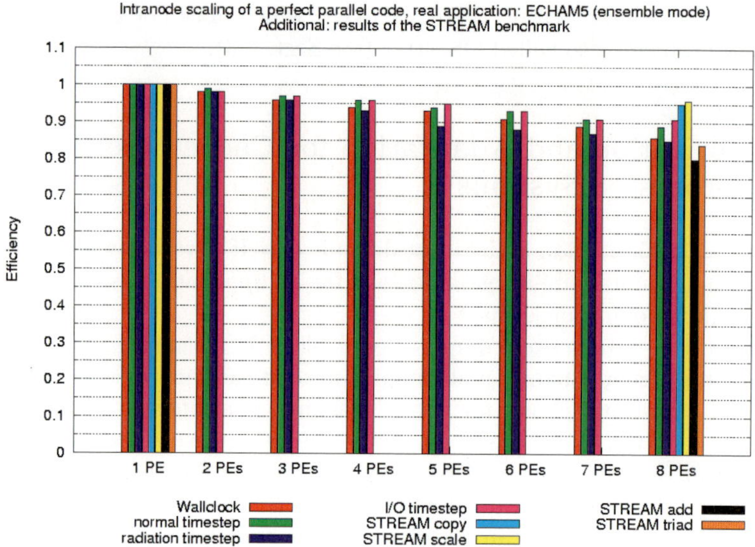

Fig. 7. Efficiency for a perfect scaling application. Comparison with some parts of the STREAM benchmark

7.2 Efficiency of the Parallel Model

When we use the model really in parallel, we have of course more effects than the systems only and this adds on to the loss in efficiency. Fig. 8 shows the degradation of the performance, increasing the number of nodes and the dependency on the models decomposition. Clearly the degradation by adding another node is quite dramatic. When assuming that the systems efficiency degradation is random the loss can be fully explained (multiplication of the efficiency on a single node with itself is a prediction for the efficiency on two nodes and so on). It looks like there is a significant impact of the system which cannot be explained by something like memory degradation, insufficient vector pipes and the parallelization of the model as shown later.

7.3 OS Jitter Effect

The effect we found is basically something like interfering applications running on a single node like the kernel, and other system processes as well as interrupts handled by different CPUs of a single node. We have been measuring that the following way. For each block of on OMP orphaned model part the execution time is measured and than summed over each thread. This is repeated several times. The results are shown in Fig. 9. Despite the large peaks

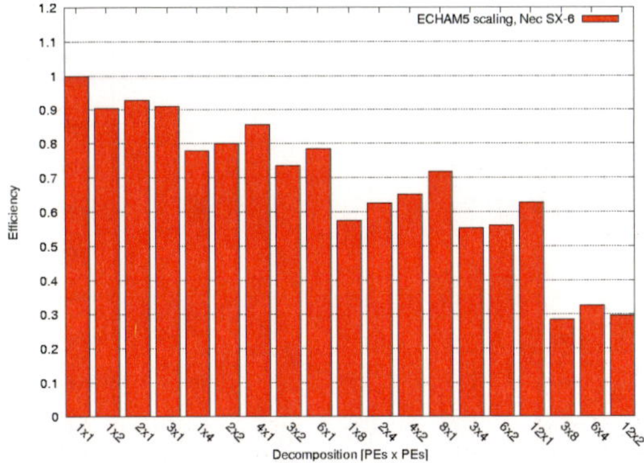

Fig. 8. Parallel scaling (efficiency) of ECHAM5 on NEC SX-6

which are caused by the additional calculation of the radiation scheme, we
see a lot of spikes. This spikes are located on different places when repeated
several times (not shown here) so that the effect is not due to any load imbalance in the model itself. By striping down a node from a production system
to a nearly empty machine the spikes could be reduced. Major impact showed
for example the process tempd.

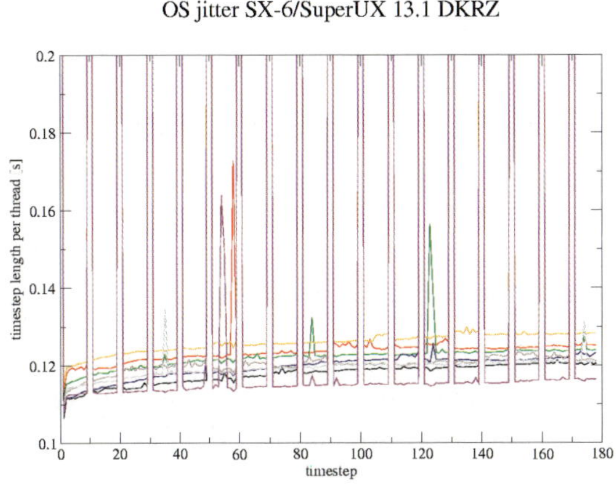

Fig. 9. Example of measurements of OS jitter

This spikes introduce load imbalance in your calculations and when your
synchronization frequency is high destroys performance and parallel efficiency.
This holds true as well on the MPI level as on the OpenMP level.

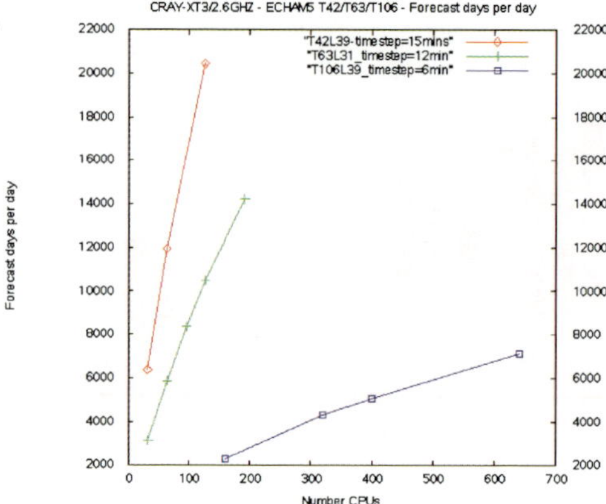

Fig. 10. Low resolution examples

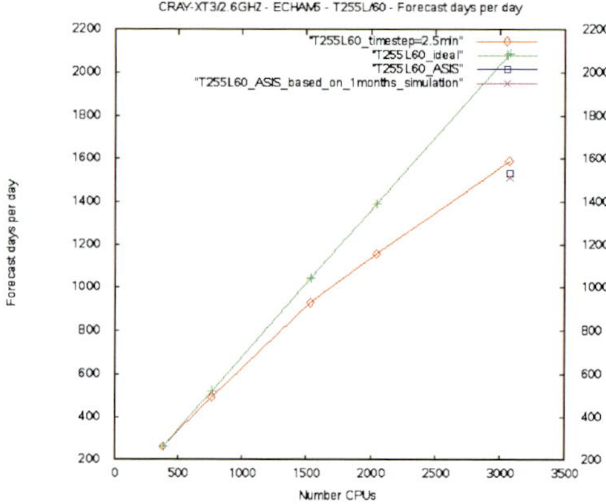

Fig. 11. High resolution example

8 Another Architecture: CRAY XT3

We got some examples (Figs. 10 and 11) of ECHAM5 scaling on an CRAY
XT3 which uses just an rudimentary micro kernel (catamount) instead of an
full OS on a single node. Here we can see that ECHAM5 can scale up to
around 3000 CPUs (still with an efficiency of 70% (Fig. 12)). This clearly
shows that its not the parallelization itself causing the bad scaling on the
NEC SX-6 system.

9 Summary

The climate model ECHAM5 is on its way to a high performance application.
Single CPU performance on vector architectures is quite good, but can be

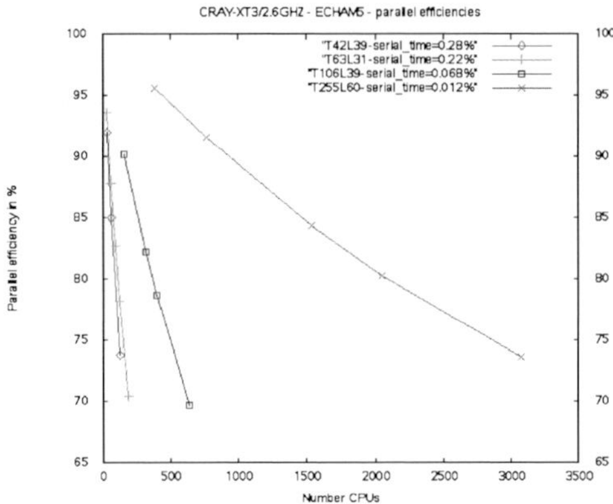

Fig. 12. Parallel efficiency

improved gradually. The OpenMP parallelization runs good on vector systems having no cache hot spots by nature of this systems. Work has to be done to allow OpenMP usage on cache-based systems as well taking care on the future multi-core developments. The MPI parallelization is very good and shows excellent scaling on appropriate build systems. OS jitter seems to have a major impact on applications with high synchronization frequency. This is a problem which has to be addressed by the design of future systems: one solution might be adding an OS CPU taking over all relevant processes. At least work in the OS scheduler has to be done to minimize the OS systems influence.

References

1. Jungclaus, J.H., M. Botzet, H. Haak, N. Keenlyside, J.-J. Luo, M. Latif, J. Maro tzke, U. Mikolajewicz, and E. Roeckner, 2005: Ocean circulation and tropical variability in the ECHAM5/MPI-OM model. J. Climate (accepted)
2. Marsland, S. J., H. Haak, J. H. Jungclaus, M. Latif, and F. Röske, 2003: The Max-Planck-Institute global ocean/sea ice model with orthogonal curvilinear coordinates. Ocean Modelling, 5, No. 2, 91-127.
3. Roeckner, E., G. Bäuml, L. Bonaventura, R. Brokopf, M. Esch, M. Giorgetta, S. Hag emann, I. Kirchner, L. Kornblueh, E. Manzini, A. Rhodin, U. Schlese, U. Schulzweida, and A. Tompkins, 2003: The atmospheric general circulation model ECHAM5. Part I: Model description. Report No. 349, Max-Planck-Institut für Meteorologie, Hamburg, 127 pp.

Printing: Krips bv, Meppel
Binding: Stürtz, Würzburg